LIFE

A Sycamore tree, *Acer pseudo-platanus*, in full foliage, showing the enormous number of leaves. After Irving.

LIFE

A Book for Elementary Students

BY

SIR ARTHUR E. SHIPLEY

G.B.E., F.R.S.

MASTER OF CHRIST'S COLLEGE
CAMBRIDGE

∴

CAMBRIDGE

AT THE UNIVERSITY PRESS

1925

CAMBRIDGE UNIVERSITY PRESS
Cambridge, New York, Melbourne, Madrid, Cape Town,
Singapore, São Paulo, Delhi, Mexico City

Cambridge University Press
The Edinburgh Building, Cambridge CB2 8RU, UK

Published in the United States of America by Cambridge University Press, New York

www.cambridge.org
Information on this title: www.cambridge.org/9781107645998

© Cambridge University Press 1925

This publication is in copyright. Subject to statutory exception
and to the provisions of relevant collective licensing agreements,
no reproduction of any part may take place without the written
permission of Cambridge University Press.

First edition 1923
Second edition 1925
First published 1925
First paperback edition 2013

A catalogue record for this publication is available from the British Library

ISBN 978-1-107-64599-8 Paperback

Cambridge University Press has no responsibility for the persistence or
accuracy of URLs for external or third-party internet websites referred to in
this publication, and does not guarantee that any content on such websites is,
or will remain, accurate or appropriate.

AMICIS

I. F. F.

A. D. H.

C. F. A. P.

OMNIBVSQVE ALIIS HVIVS OPVSCVLI
FAVTORIBVS ATQVE ADIVTORIBVS
D. D.

GRATO ANIMO

A. E. S.

AMICVS

PREFACE

A YEAR ago the University Press asked me to write a book which would make students of elementary Biology think. I do not know in the least whether I have succeeded in doing so. The average schoolboy, especially at the age when he usually begins to study Biology, is strongly of the opinion that "thinking is but an idle waste of thought," and with few exceptions he turns away from the advice of one of the wisest and worldliest of our teachers. "Of all the truths do not decline that of thinking. The host of mankind can hardly be said to think," as Lord Chesterfield wrote to his son.

What I have tried to do in this book is to emphasize the unity of life, whether it be plant-life or animal-life, and the interrelation of living organisms one with another and with their surroundings. The crayfish with its *scaphognathites* and *dactylopodites*, and the fresh-water mussel with its *ctenidia* and its *osphradia* do not live self-contained lives tucked away in water-tight compartments. They are in intimate relation with the whole world of other plants and animals and with their physical surroundings. The dead dogfish in a dissecting dish gives one but little idea of what it did and of what happened to it when it was alive. I have tried to bring out the fact that plants and animals are at one in being alive, and I have tried to make clear the intimate association of both with their environment, whether it be the air or the soil or the sea. The whole of life is so interwoven and interconnected that the "type-system," however well it may teach us the rudiments of Anatomy, gives a totally inadequate representation of life.

PREFACE

It is easier to examine live plants in a laboratory than live animals. They are easier to rear. They reproduce as a rule more quickly and more continuously and they require much less attention. Further, all the older Universities of Europe and many of our Schools have Botanical Gardens, but I have never come across a University or a School which has an adequate Zoological Garden.

The only really difficult part of this book is that dealing with the alternation of generations in plants. But it is such a wonderful illustration of evolution and its whole story is so romantic that I have ventured to give a short sketch of its progress from the simplest seaweeds to the highest flowering plants, a sketch far too condensed. Here at least the student may require the aid and explanations of a teacher.

Finally I venture to hope that this book will be not without interest to the public that is not preparing for examinations, and thank heaven that public is still in the great majority!

A. E. SHIPLEY.

CHRIST'S COLLEGE LODGE, CAMBRIDGE.
    17th November, 1923.

PREFACE TO THE SECOND EDITION

There are comparatively few alterations in this new edition of "Life." A few crooked paths have been straightened out, but the most important alteration is the re-writing of page 36. In the first edition the account of the connexion between chlorophyll and haemoglobin was not accurate.

I have also incorporated a more modern view of the function of the contractile vacuole.

For many of the corrections I am indebted to my friends Mr J. Barcroft, F.R.S. of King's College and Mr C. F. A. Pantin of the Marine Biological Laboratory, Plymouth.

A. E. S.

CHRIST'S COLLEGE LODGE.
    4th February, 1925.

CONTENTS

Chapter	Page
I. INTRODUCTION	
Definitions of Life	1
Protoplasm	2
Attributes of Living Matter	3
II. PROTOPLASM AND CELLS	
The Constitution of Protoplasm	7
The Amoeba	9
Tissues	13
III. CELLS AND THEIR PARTS	
Cells	14
Cilia and Flagella	15
Plant-Cells	17
Colonies	19
Individuals	20
IV. FEEDING	
The Feeding of Plants and Animals	21
Types of Feeding	23
Parasites	24
Diatoms	25
Bacteria	26
V. CHLOROPHYLL	
Chlorophyll	28
The Oxygen and Carbon Dioxide Cycle	29
Mineral Oil and Natural Gas	34
Haemoglobin	36

CONTENTS

Chapter		Page
VI.	THE NITROGEN CYCLE	
	Sources of Nitrogen	37
	The Protoplasm Cycle	41
VII.	THE SOIL AND THE SAP	
	Composition of the Soil. Leaf Mould	43
	Life in the Soil	45
	The Sap	48
VIII.	FOOD	
	Chemistry of Food	57
	Variety of Foodstuffs and Feeding Habits	58
IX.	DIGESTION	
	Alimentary Canal	87
	Body-cavity	88
	Digestion	90
	Hormones and Vitamins	92
	Calories	95
	Appetite	96
X.	RESPIRATION	
	Plant Respiration	99
	Respiratory Organs in Animals	101
	Dust and Sand	105
	Haemoglobin	106
	Anaerobes	112
XI.	MOVEMENT	
	Movement of Plants	115
	Amoeboid Movement	121
	Movement of Animals	122
	Flight	128

CONTENTS

Chapter		Page
XII.	RHYTHM	
	Rhythm in Parts of Cells.	133
	Rhythm in Cells	134
	Rhythm in Tissues.	136
	Rhythm in Organs.	138
	Rhythm in Organisms	142
	Rhythm in Communities.	154
XIII.	REPRODUCTION	
	Growth and Reproduction	160
	Spores	161
	Vegetative Reproduction.	162
	Sexual Reproduction	165
	Alternation of Generations	167
	Number of Offspring	175
	Origin of Life from the Sea	179
	The Egg.	182
	Antherozoids and Spermatozoa	183
	Hermaphroditism	185
	Parthenogenesis	186
	Old Age and Death.	191
	INDEX	194

ILLUSTRATIONS

Figure		Page
	A Sycamore tree, *Acer pseudo-platanus*	*Frontispiece*
1.	Various stages of a Myxomycete, *Chondrioderma difforme* From Strasburger.	9
2.	*Amoeba proteus* From Gruber.	11
3.	A shell of *Globigerina* After Brady.	14
4.	*Euglena viridis*	15
5.	A leaf cell, highly magnified From Vines.	18
6.	*Volvox aureus* Ehrenb..	19
7.	A cell of *Spirogyra* Darwin's *Elements of Botany*.	29
8.	Erect shoot of Norway Maple, *Acer platanoides* Kerner.	30
9.	Surface view and section of leaf of London Pride Thoday's *Botany for Senior Students*.	32
10.	Root of Broad Bean, with nodules After Strasburger.	38
11.	Root-hairs Thoday's *Botany for Senior Students*.	48
12.	Seedlings of Mustard After Sachs.	49
13.	Sieve-tubes and vessels from the stem of a Sunflower Thoday's *Botany for Senior Students*.	50
14.	Leaf of Plane, *Platanus orientalis* From Ettingshausen.	51
15.	Semi-diagrammatic view of a leaf in section	53
16.	Three stomata with surrounding epidermic cells	54
17.	Transverse section through the leaf of the Hellebore	55
18.	The Medicinal Leech, *Hirudo medicinalis*	60
19.	Underview of an Indian Scorpion, *Scorpio swammerdami* Shipley and MacBride's *Zoology*.	66

ILLUSTRATIONS

Figure		Page
20.	The Warty Newt, *Malge cristata*. From Gadow.	68
21.	The Texas Rattlesnake, *Crotalus atrox*. From Stejneger.	70
22.	The Duckbill, *Ornithorhynchus anatinus*	72
23.	The Rock Wallaby with young in pouch, *Petrogale xanthopus* After Vogt and Specht.	73
24.	Tamandua Ant-eater, *Tamandua tetradactyla* From Proc. Zool. Soc. 1871.	75
25.	The Six-banded Armadillo, *Dasypus sexcinctus* After Vogt and Specht.	75
26.	Indian Rhinoceros, *Rhinoceros unicornis* From Wolf.	78
27.	The Musk-Ox, *Ovibos moschatus*	80
28.	The Musquash, *Fiber zibethicus*	81
29.	The Common Skunk, *Mephitis mephitica*	82
30.	Russian Desman, *Myogale moschata*	84
31.	Female and young of *Xantharpyia collaris*. From Sclater.	85
32.	The Orang-utan, *Simia satyrus* From a specimen in the Cambridge Museum.	86
33.	Diagram of digestive and excretory system of the Liver-fluke, *Distomum hepaticum* From Leuckart.	88
34.	A Sea-lily, *Antedon acoela* After Carpenter.	106
35.	Horizontal underground stem, or rhizome of *Pteris* Darwin's *Elements of Botany*.	115
36.	Horizontal underground stem, or rhizome of a sedge. From Le Maout and Decaisne.	116
37.	Stages in the collapse of a cell when immersed in a strong solution After De Vries, modified.	117
38.	*A*, Hop twining with the sun; *B*, Convolvulus twining against the sun. After Payer.	118

ILLUSTRATIONS

Figure		Page
39.	Seedling of the Bean, *Vicia Faba*. Darwin's *Elements of Botany*.	119
40.	A Laurel twig, *A* in the frozen, and *B* in the thawed condition Darwin and Acton's *Practical Physiology of Plants*.	120
41.	*Paramoecium caudatum* After Bütschli.	123
42.	*Planaria polychroa* swimming	124
43.	The Edible Snail, *Helix pomatia*. From Hatschek and Cori.	124
44.	Underview of a Starfish, *Echinaster sentus* From Agassiz.	126
45.	Map showing the range of the American Golden Plover From the Bureau of Biological Survey, Washington.	130
46.	Transverse section of an Oak-trunk From Le Maout and Decaisne.	136
47.	Otoliths of Plaice. After Wallace.	138
48.	Section of a Pearl From Rubbel.	139
49.	Shells of a Fresh-water Mussel, *Anodonta mutabilis*	139
50.	*Iulus terrestris* From Koch.	140
51.	*Nereis pelagica* After Oersted.	145
52.	View of female *Cyclops* sp. After Hartog.	152
53.	The Wood-ant, *Formica rufa*	155
54.	The Honey-bee, *Apis mellifica*	157
55.	A Wasp, *Polistes tepidus*	158
56.	Yeast highly magnified Darwin's *Elements of Botany*.	161
57.	Creeping Buttercup (Praeger's *Weeds*.)	163
58.	*Bougainvillea fructuosa*. From Allman.	168

ILLUSTRATIONS

Figure		Page
59.	Vertical section through a fertile leaf of a Fern, *Nephrodium* After Krug.	170
60.	Young plant of Maidenhair Fern still attached to prothallus. After Sachs.	171
61.	Sexual organs of a Fern, *Polypodium vulgare*	171
62.	Longitudinal section through a nucellus After Darwin.	173
63.	Pollen-grains germinating on the stigma of a Grass After Kerner.	174
64.	Ovary of *Polygonum convolvulus* during fertilization After Strasburger.	175
65.	A Dogfish and egg, *Scyllium canicula* From Day.	176
66.	*Amphioxus lanceolatus* After Lancaster.	178
67.	A Lung-fish, *Ceratodus forsteri* After E. S. Goodrich.	179
68.	Tadpoles Latter's *Natural History of some Common Animals*.	180
69.	Side view of *Cypris candida* After Zenker.	184
70.	The Apple Aphis, *Aphis pomi* Carpenter's *Life Story of Insects*.	188

LIFE

CHAPTER I

INTRODUCTION

DEFINITIONS OF LIFE—PROTOPLASM —ATTRIBUTES OF LIVING MATTER

"Sairey," says Mrs Harris, "sech is life. Vich likeways is the hend of all things."

Mrs Gamp. *Martin Chuzzlewit*. CHARLES DICKENS.

DEFINITIONS OF LIFE

At the Dundee meeting of the British Association for the Advancement of Science held in 1912, the President introduced the subject of Life, and the topic proved undoubtedly interesting and even stimulating. It led to much discussion, but, as far as I am aware, no one tried to define life or even threw much new light on the question the savants so eagerly discussed:

> Myself when young did eagerly frequent
> Doctor and Saint and heard great argument
> About it and about; but evermore
> Came out by the same door as in I went.

This seems to be the fate of anyone who tries to define life. *The Oxford Dictionary* tells us that life is "the condition or attribute of living or being alive; animate existence. Opposed to *death*." This definition at once begs the question and argues in a circle. Dr Johnson takes a more eighteenth century attitude and says life is "union and co-operation of soul with body; vitality; animation, opposed to an *inanimate state*." One would like to have heard Dr Johnson's opinion on protoplasm! Even Herbert Spencer's formula that life is "the continuous adjustment of internal relations to external relations..." omits the fundamental consideration that we know life only as a quality of

INTRODUCTION

and in association with living matter. When I was at school they used to tell me that a verb indicated "being, doing or suffering" and this certainly describes "life," though it does not define it. "Och, life's aye a laugh and a greet," as Sir Harry Lauder tells us. And then there's the well-known character who defined life as "one d—d thing after another"; but he was referring to a span of life—"Brief life is here our portion." It was probably the same pessimist who said, "The moment you're born, you're done for." For

> Life lives on death; Death lives on life;
> And so the circle runs,
> Throughout a million teeming years,
> And half a million suns.

Perhaps the best way to describe life is to enumerate those qualities which living organisms have and non-living objects have not, and then to say that life is the expression of these qualities. In the "Introductio" to his *Philosophia botanica* (1751) Linnaeus, who had named more plants and animals than anyone since Adam, states: *Lapides crescunt. Vegetabilia crescunt & vivunt. Animalia crescunt, vivunt, & sentiunt*, and roughly speaking this is true.

Protoplasm

Living matter or, as Huxley phrased it, the "physical basis of life," is a substance called by Hugo von Mohl *protoplasm*. Since this protoplasm is always being added to from the outside world in the form of food and oxygen, and *per contra* is always giving up something to the outside world in the form of carbon dioxide breathed out and of other excreta, some might regard it more in the light of a space, in which various elements enter, combine, disintegrate and take their exit, than as a substance. Still, for the convenience of this book we will regard it as a substance never constant in composition for a single second.

To see this protoplasm in any mass and to form some idea of what the substance looks like it is better to study some of the larger of those animals which have but a single cell, or the contents of some of the larger cells among the plants, for

PROTOPLASM

here we can look at it, under the microscope, undifferentiated and, so to say, in bulk. If we do so look, we see a whitish substance, sometimes clear as crystal, but more often semi-opaque, like ground-glass. It contains many darker specks or granules, and some of these are particles of food. If the protoplasm be shut up in a vegetable-cell it flows hither and thither, usually up one side and down the other, or, like a Roger de Coverley dance, "up the side and down the middle." If the protoplasm be free, *i.e.* not confined by any surrounding cell-wall, it will be constantly changing its outline, on one side thrusting out a lobe or protuberance, on the other perhaps withdrawing one, and in this way the whole piece of protoplasm may move slowly forward. This whitish, soft substance, semi-jelly, semi-fluid, is living protoplasm, but so are our muscle cells and the cells of our brain and our blood corpuscles. All these, however, are more specialized and not so easily studied; still, all obey the same laws and do the same ultimate things.

ATTRIBUTES OF LIVING MATTER

What is it that this protoplasm does that non-living matter, such as rocks and stones, never does? To begin with, it is *motile*. We have seen that it can alter from time to time its outline or shape, and by doing this in a certain way it can move forward or progress, or move backward and regress. Therefore it is *motile*, and the slow protuberance of a lobe on one side of the body, and the equally slow withdrawal of another on the other side, is the first beginning of that muscular contraction which may ultimately produce a competitor for the Olympic Games. As far as we can judge, even this simple movement is not always the result of an external stimulus, but arises from something in the protoplasm itself, and certainly such is the case in the more complex instances of higher life. This initiation of action from within is called *automatism*, and protoplasm, unlike non-living matter, is *automatic*. But it also readily reacts to external impressions or stimuli. An electric current passed through the water in which our living matter is suspended will cause it to contract into a sphere, and thus to present for its bulk the smallest possible surface to the stimulus of the current; or, again, a piece of food will attract it, and towards that piece of food it will slowly move—in short it responds to external stimuli, and is, as the physiologists call it, *irritable*.

These activities and qualities imply a certain expenditure of energy; how is that energy supplied? What is the oil that drives this engine? It is the food already hinted at. Living protoplasm must have food. It takes to itself certain food substances of a high complexity and oxidizes and reduces these to simpler substances, and during this process, just as when gunpowder explodes, energy and heat are set free. It is also capable of building up the dead food into its own flesh (or into protoplasm), making the dead live, and this quality is called *assimilation*. Further, all protoplasm breathes; that is to say, it takes in oxygen and it gives out carbon dioxide. It is in effect *respiratory*. Should the supply of oxygen in this world of ours be suddenly withdrawn, all life would cease, and in the course of a few weeks or months the whole fabric of our earth would have become mineralized. Life would cease.

Living animals and plants *secrete*, that is to say, certain cells and organs bring forth products which play a large part in the life of the animal or plant in question. Examples of such glands or secretory cells in plants are found in the nectaries of flowers, which secrete a sugary fluid attractive to insects, who bring with them pollen from other plants to fertilize the flower. The glands of insectivorous plants produce a fluid which digests the proteins of the insects they catch, and other cells secrete certain solvents which render the starch in seeds soluble.

In the higher animals the products of the secretory glands play a large part in digestion. The salivary glands secrete saliva into the mouth, which moistens the food and turns starch into sugar. The walls of the stomach secrete gastric juice, which together with the products of the liver and the pancreas help to render undigested food soluble so that it may be taken up into the blood-stream. Then there are certain small glands which have no ducts and do not open anywhere. But their secretion passes into the blood and plays a large

EXCRETION—REPRODUCTION—RHYTHM

part in various functions of the body, such as growth, and they are also the cause of many obscure diseases. A secretion is a product which serves a useful purpose in the economy of the organism.

But with *excretion* we have to deal with the formation of bodies which are useless or injurious to plant or animal. Unless these are discharged from the body, the whole organism gets choked, just as ashes may put out, in time, the driving fire of a steam engine. Plants as a rule store away their excreta in parts of the body where it is harmless, although those plants that cast their bark and shed their leaves get rid of a good deal of excreta annually. Animals excrete sweat, urea, and certain products from the alimentary canal, but the great bulk of the last-named have never formed part of the organism which rejects them. They have passed through as undigestible portions of the food. In certain animals such as the ASCIDIAN, or seasquirt, the urea is stored away where it is harmless to the body, but in most animals the urea is taken up from the blood by the kidneys and passed to the exterior. Carbon dioxide (CO_2) is another excretion which passes away from the gills or lungs, and out from the skin like the sweat. A certain amount of excreta is got rid of by ARTHROPODA (CRUSTACEA, INSECTA, etc.) when they cast their skin, and the same is true to a certain extent of some worms. Matter which is not living does not secrete or excrete.

Living matter grows and reproduces. Animals and plants give rise to successors and they, in their turn, reproduce. The most primitive method of reproduction is that the animal or plant splits or cleaves in two. Each of the two will then increase in size until it reaches a certain bulk—this is *growth*—and then again it will divide. No dead or non-living object behaves in this way.

Finally, living matter is *rhythmic*. It is always doing something or other at stated intervals. These intervals often seem to have no relation to outside influences, like breathing or the recurrent beats of a heart; but in many cases the intervals between the acts correspond with cosmic changes. Night and day control sleep; the tides have a marked influence on the habits of many of the shore-living Invertebrates, and so ingrained are these periodic habits that they are retained even when the animal showing them is removed inland and kept in a perfectly still aquarium. Summer and winter, seedtime and harvest play perhaps the greatest rôle in this rhythm. One has only to think of the breeding habits of most animals or of the annual appearance and disappearance of the foliage of deciduous trees to recognise this.

Yes; life, if undefinable, is rhythmic.

CHAPTER II

PROTOPLASM AND CELLS

THE CONSTITUTION OF PROTOPLASM
—THE AMOEBA—TISSUES

The remainder of the cell is more or less densely filled with an opaque, viscid fluid, of a white colour, having granules intermingled in it, which fluid I call protoplasm.
                                                    VON MOHL (1846).

THE CONSTITUTION OF PROTOPLASM

ALL living organisms are built up of protoplasm and its products. Both plants and animals consist of this same protoplasm, the "physical basis of life," as Huxley called it. All modern evidence tends to show that protoplasm is an equilibrium mixture of a fluid and of a more solid jelly. The relative proportions of the liquid and the jelly are from time to time changed. Many cells are solidified at certain times to a high degree; and the change from the more fluid state to the more jellied state is reversible. Cells which are at one time very fluid may at other times be very solid, and *vice versa*. For instance, fertilization causes the protoplasm of the egg to become less solid and more liquid.

It is impossible to analyse by chemical or physical means *living* protoplasm, for any attempt at such analysis at once kills it.

By the analysis of dead protoplasm we find it contains *proteins*, and proteins are compound chemical substances which are never found apart from living matter. They contain carbon, hydrogen, nitrogen, oxygen and sulphur, and have the following percentage composition:

Carbon	from 50	to 55	per cent.
Hydrogen	,,	6·5 ,,	7·3	,,
Nitrogen	,,	15 ,,	17·6	,,
Oxygen	,,	19 ,,	24	,,
Sulphur	,,	0·3 ,,	2·4	,,

The proteins are tremendously important, and play a most prominent part in the building up of protoplasm. Proteins are never absent from living matter, and except for the fact that some of the simpler kinds can be synthetically produced in the laboratory they are never derived from any other matter than that which is living. The building up of proteins is the most important factor in life.

Proteins are highly complex, and are very varied in their nature; but all react in the same way to certain chemicals. Proteins in food differ from those in the living tissues. During digestion the former are broken up and then reconstructed, only to be finally broken down again into carbon dioxide, water, sulphuric acid, urea, and other products which are excreted from the body. Animals convert the protein of their vegetable food into the protein of their own body. But the building up of proteins in plants involves the combination of the soluble nitrogenous food taken up by the root with the complex carbon compounds formed by the green leaves in sunlight, and we thus have two circles or cycles, the cycle of carbon, which will be described in Chapter V under the title of Chlorophyll, and the cycle of nitrogen, which will be described in Chapter VI under the heading of The Nitrogen Cycle.

The molecules of the proteins are very large, certainly the largest and most complex that are known; and consisting as they do of thousands or even tens of thousands of atoms they afford a large scope for the slight variations which we find in the different classes of proteins. The variety in the arrangement of these numerous atoms in the complicated molecule may also explain the differences which exist between the different species of plants and of animals. The protoplasm of one species of animal or plant differs from the protoplasm of all others. "All flesh is not the same flesh; but there is one kind of flesh of men, another flesh of beasts, another of fishes, and another of birds."

It may be that differences between plant and plant and animal and animal depend on minute differences in the structure of their proteins, or possibly between the ways in which the protein molecule is built up. There are also minute differences, which are nevertheless perceptible, between the starches of one plant and the starches of another. There are

PROTOPLASM—THE AMOEBA

similar differences between the haemoglobins of red-blooded animals and these differences are to some extent characteristic of the species concerned.

Other elements than those found in proteins also occur in the bodies of plants and animals. Phosphorus, chlorine, potassium, sodium, magnesium, calcium and iron are all found, having been taken in with the food or water.

Further, living protoplasm has not a constant composition. It is changing every moment, taking new matter into itself and discharging other matter. It is, in fact, like a stream or flame. Six centuries before Christ, Buddha, in his last reincarnation, maintained that "Life is a flame," and, like a flame, protoplasm is never the same but always changing its composition. It is a living example of the saying of Alphonse Karr about the French Government under Louis Philippe: "Plus ça change plus c'est la même chose."

The Amoeba

To form an idea of what protoplasm *looks* like one might examine with a microscope the uncooked white, *albumen*, of an egg, or a drop of fairly thin gum. Both are glairy, semi-transparent, full of particles: but neither of them is protoplasm. Perhaps the largest masses of more or less undifferentiated protoplasm easily visible to the naked eye are those curious slime-fungi, MYXOMYCETES, which are found several inches in diameter slithering about on dead and rotten wood in damp forests. But one cannot always find slime-fungi, and a better plan is to examine under the microscope the unicellular organism known as *Amoeba*. *Amoebae* are common enough in both fresh and salt water and in the soil. They are irregular in shape and their outline is

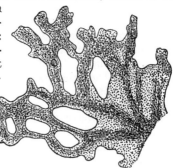

FIG. 1. A Myxomycete, *Chondrioderma difforme.* From Strasburger.

constantly changing. Their average diameter is about 100–250 μ[1].

Lobes, *pseudopodia*, are constantly being thrown out from any part of the surface of the body and then withdrawn. The very thin outer surface coating of the minute creature is clear and free of granules, but the great bulk of the animal is slightly opaque and is full of granules of various sizes, some of which are food particles. The *Amoeba* will slowly crawl towards any small organism and around this it will thrust a couple of lobes or arms, forming a bay. The tips of the arms will unite, and then we have a small lake, in the centre of which the engulfed food-particle is now floating. Digestive fluids are passed from the protoplasm into this food-vacuole, and when the food is digested the refuse passes out of the *Amoeba* by a reverse process. The fluid surrounding the engulfed particle is at first acid and later it becomes alkaline, just as the contents of our stomach is acid and that of our intestine alkaline. This fluid dissolves the food so that it can be incorporated in the surrounding protoplasm. The pseudopodia of an *Amoeba* at times exert a surprisingly great force. They are even capable of nipping a *Paramoecium* in two, engulfing one half and leaving the other half outside, and a *Paramoecium* is a pretty tough organism.

Embedded somewhere near the centre of the little animal is a more solid ball which takes up stains more freely than the rest of the protoplasm; this is the *nucleus*. During life the nucleus is almost invisible, and it is larger in active, busy cells than in quiescent, inactive cells. Its functions include both the control of the building up of the food into protoplasm, *assimilation*, and the control of reproduction. If an *Amoeba* be cut in two, one half with, the other without, the nucleus, the nucleated fragment behaves as a normal *Amoeba*. The non-nucleated fragment also behaves quite normally *except* that it cannot divide and cannot digest its food, though food particles may be ingested. The life of such a non-nucleated

---

[1] A μ is the thousandth part of a millimetre. The smallest thing you can see with an ordinary microscope must have a diameter of 0·14μ and with an ultra-microscope about 0·005μ. The smallest bacterium which has yet been seen has a diameter of about 0·5μ.

THE AMOEBA

Amoeba ends in death after a period of time about the same as that of a normal *Amoeba* which has been starved. This indicates that the function of the nucleus is to control metabolism and reproduction rather than to maintain the activities of the *Amoeba* as a whole.

Usually one large *water-* or *contractile-vacuole* can be seen, which from time to time contracts and squirts a liquid out of the body. This is probably an osmotic organ controlling the

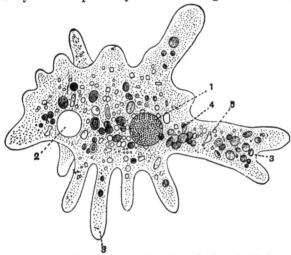

FIG. 2. *Amoeba proteus*. ×330. From Gruber. 1. Nucleus. 2. Contractile vacuole. 3. Pseudopodia; the dotted line points to the clear ectoplasm. 4. Food vacuoles. 5. Grains of sand.

amount of water which has passed by osmosis into the saline protoplasm. Without such an organ the cell might burst. The waste products are got rid of from the general surface of the cell, for if they accumulate they interfere with the life of the *Amoeba*, just as too many ashes put out a fire.

A recent device enables us to dissect organisms as small as *Amoebae*, under the microscope. Exceedingly fine glass needles are used. The contents of the contractile vacuole if stirred up with the surrounding protoplasm by means of such a fine needle cause a very rapid and complete break-up, *histolysis*, of the entire animal; this shows how essential is the removal of the waste products.

The *Amoeba* moves, and it moves of its own accord. It will select one out of several particles of food and move toward it. Thus it is largely *automatic* in its responses.

But it is also irritable. If you pass an electric current through the fluid containing an *Amoeba* on a microscopic slide it will contract into a sphere, thus presenting the least possible surface to the irritant; or it will move away from a chemical that annoys it. But it is difficult to deny that this movement may be a response to some chemical action on the part of the food. *Amoebae* floating in distilled water and white blood corpuscles, which are very like *Amoebae*, floating in the liquid of the blood, out of contact with all solid bodies, are said not to throw out pseudopodia. If this be so, it is probably due to the fact that there is an absence of salts in the distilled water. Salts certainly have a marked effect in the production of pseudopodia. Many *Amoebae* when floating freely in a clear medium throw out a number of fine active radiating pseudopodia. The question of "free-will" in an *Amoeba* may still be a subject for argument. Still, there is evidence that a certain amount of discrimination can be exercised even by the lowest, unicellular animals.

The *Amoeba*, like nearly all animals and plants, *respires*, that is it takes in oxygen and gives out carbon dioxide, and this it does and they do "all the time." Certain sugars in the cell are oxidized or broken up, potential energy becomes kinetic energy and this takes the form of heat and movement. The chemical result of this reaction or breaking up of the complex molecule is carbon dioxide and water, and these leave the body of the breathing organism by gills, or lungs, or sweat glands, or kidneys, or, as in the *Amoeba*, by the general surface of the body.

Protoplasm is *reproductive*, and the *Amoeba* exhibits the simplest form of reproduction. First of all, the nucleus divides. Then a waist appears in the *Amoeba* between the two nuclei; this waist becomes smaller and smaller, but as long as it is there the protoplasm flows through it from side to side. But after a time it snaps, and we have now two

Amoebae where before we had only one; and no doubt if they have a good conceit of themselves, each thinks itself the mother of the other. Division follows normally only on the growth of an *Amoeba* to a certain size, which is larger than the average. Well fed cultures divide more rapidly than poorly fed ones.

Many different species of *Amoeba* have been recorded, and their life-history, which is often more complex than that described above, has been studied; but there is always a "snag" somewhere, and many organisms that look like *Amoebae* are simple amoeboid stages in the life-history of higher organisms. Thus the freshwater *Hydra* has an egg which is indistinguishable in structure from the typical *Amoeba*. Only by breeding and rearing in cultures can the real nature of these amoeboid organisms be determined.

Tissues

We have seen that the *Amoeba* is a single cell, and this single cell can perform many functions. It can feed and breathe and move and excrete and reproduce; but in higher multicellular organisms these functions are assigned to, and carried on by, cells especially adapted to these ends. In a primitive state of society, amongst savages, each man is his own hunter, fighter, butcher, baker, builder, gardener, and, in fact, universal provider; and with the exception that he cannot be his own undertaker he manages to carry out all the functions which in higher and more civilized societies become the work of specialized or expert craftsmen. The latter is the case with organisms that consist of many cells—the *Metaphyta*, multi-cellular plants, and *Metazoa*, multi-cellular animals. There are muscular cells which bring about movement, nerve cells which are responsible for irritability and automatism, cells for digestion, cells for breathing, and reproductive cells. Accumulations of cells, similar in structure and in function (*i.e.* doing the same things), are called *tissues*. Thus we have storage-tissues in plants, muscular or nervous tissues in animals and so on. The tissues are built up into organs such as roots and leaves in plants, or the heart and brain in higher animals, and the organs are again built up to form with others the body of the plant or animal.

CHAPTER III

CELLS AND THEIR PARTS

CELLS—CILIA AND FLAGELLA—PLANT-CELLS —COLONIES—INDIVIDUALS

"Wherefore one should not be childishly contemptuous of the most insignificant animals. For there is something marvellous in all natural objects." ARISTOTLE.

CELLS

THE *Amoeba*, as we have seen above, is one of the simplest of those animals which consist of one single *cell*, but there are an incredible number of different forms of unicellular animals, many of them of the greatest complexity and of an infinite variety of shape. Some of them secrete skeletons which may be either of chalk or of flint. If you pound up, not too finely, a piece of chalk and examine it under a microscope, you will see a number of little chalky shells, some of them of many chambers, which were all secreted by a single cell with many nuclei or one. The immense quantity of

 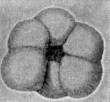

FIG. 3. A shell of *Globigerina*. After Brady.

these little calcareous shells is shown by the fact that some of our Chalk measures, *e.g.* those of Norfolk, have an average thickness of some 1450 feet, and the huge areas of the bottom of the ocean are covered with such minute shells, forming a greyish-brown paste, known as *Globigerina* ooze. Nearly fifty million square miles of the whole surface of the bottom of the deep sea is covered with such chalky shells. Another 2,290,000 square miles of the ocean bed is floored with flinty shells formed by other unicellular organisms. The

CELLS AND CILIA

flinty shells are commonest in the deeper parts of the ocean, which in some cases is five miles below the surface of the water. It is quite impossible to put into any kind of figures that can be realized the incredible numbers of these minute organisms which have built up and are building up so large a portion of the earth's crust; even a Russian bank-manager used to the Soviet currency would hardly grasp their significance.

As a rule, animals with one cell are not visible to the naked eye, but there are certain unicellular animals that are parasitic in the bodies of higher animals which are clearly to be seen without magnification. A parasite is an animal that lives in another plant or animal—called the "host"—and obtains its nutriment at the expense of the host. Certain unicellular parasitic animals called *Gregarines*, which live in earthworms, are quite visible to the naked eye, and one which lives in the lobster attains a length of two-thirds of an inch.

CILIA AND FLAGELLA

Amoebae move by a creeping or gliding action, but many unicellular plants or animals are propelled or rowed along by vibrating processes called *flagella*. When a cell has a number of flagella all vibrating in unison, they are called *cilia*. Cilia play so large a part in life, not only in Protozoa but in most of the higher animals, that they deserve a special section of this chapter. Cilia are like eye-lashes standing out from the body of a cell. There may be only one, as is the case with many unicellular plants or animals or the male reproductive cells of plants or animals, the *antherozoids* and the *spermatozoa*, and then, as we have said, it is called a flagellum. Such a flagellum may be at the front end or the hind end of the cell. In the former case it lashes about, sometimes in a spiral, and draws the cell after it. Or it may

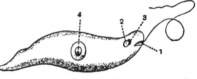

FIG. 4. *Euglena viridis*, which swims by the lashing of a single flagellum at the front end. ×100. 1. Mouth. 2. Contractile vacuole. 3. Pigment spot or eye. 4. Nucleus.

be at the hinder end of the cell, in which case it pushes the cell forward like a propeller in the stern of a boat. But much more commonly each cell has a group of cilia which wave to and fro, as the stems in a field of corn wave when swept by a wind and right themselves when the wind has passed. Such flickerings row an isolated cell along like the rhythmic beat of the oars in a boat.

In higher multi-cellular bodies, like our own, cilia are used to push bodies along. They move to and fro, creating a current of fluid or mucus, and in this current particles are swept forward. In the human body the air passages are lined with cilia, and these cilia drive bacteria and other foreign bodies up towards the mouth and so in time out of the body. They also line the oviducts, and help to push the human egg down into the uterus. Others have been found in the testis, and again ciliated cells line the cavity of the brain and the minute central canal which passes down the spinal cord. The cilia which move the spermatozoa in search of the ova may also be regarded as flagella. In the gullet of the frog cilia work downwards and help the creature to swallow. In many aquatic animals, such as the oyster and the mussel, which are either fixed or very sluggish, they keep the body supplied with fresh volumes of water which bring food to the mouth of the mollusc and oxygen to its gills, and the same holds good in many other fixed, *sessile*, aquatic animals. Some members of the freshwater animals, known as Rotifers or wheel-animalcules, do not move about, and here cilia create a current which is the sole source of the food-supply of these curious little creatures.

The cilia of an isolated cell from a frog's gullet, floating in salt solution and completely removed from all nervous control, will continue to beat rhythmically for many hours, yet in the body of the frog their beat may be slowed down or completely stopped by various stimuli. The beating of cilia from the mantle of certain molluscs stops at once when stimulated by an electric current; however, if the mantle is first treated with an anaesthetic such as novocaine, the electric stimulus produces no effect and the cilia continue to beat. It is believed that the first, or primary, inhibition, or stopping

PLANT-CELLS

of the beating, is brought about by some nervous mechanism stimulated by the electric current, and the anaesthetic is believed to do away with this nervous mechanism. Electric currents which would cause the immediate dissolution, *histolysis*, of a host of protozoa apparently have no effect upon the mantle of the mollusc or its cilia after treatment with anaesthetics.

Cilia play so large a part in the life of higher animals that it is a very extraordinary fact that two great groups of multi-cellular animals are entirely devoid of them. The Round-Worms or NEMATODA, which are frequently parasitic in both higher plants and animals, are devoid of cilia; and so is an enormous group of animals which reaches the highest grade of complexity and social organization, known as the ARTHROPODA. There are no cilia to be found in shrimps, lobsters, crabs, in centipedes, spiders, mites or in insects. When one considers the immense part cilia play in the life of all the other great groups of animals one wonders that this great and predominant class can get on without them. Cilia and flagella are not so common in plants, but they exist in the male reproductive cells in sea-weeds, mosses, ferns and even in some primitive trees.

PLANT-CELLS

As a rule plant-cells are enclosed in a firm cell-wall, and the name "cell" originates from the fact that when the microscope was first discovered small slices of cork were investigated, and cork consists of a series of the cases of dead cells which are as regularly arranged as the cells in a honeycomb. The living matter or protoplasm of plant-cells is thus not naked, as it is in so many animal cells, but is enclosed, as it were, in a suit of mail armour. But the protoplasm of one cell communicates by extraordinarily minute channels with the protoplasm of neighbouring cells, and further, this case or coating does not prevent the protoplasm from moving. It often rotates within the cell-wall; it may be flowing up both sides of the cell and returning down the middle like a country dance, carrying with it various food granules and at times the nucleus.

18 CELLS AND THEIR PARTS

Owing to the plant-cell having a wall of *cellulose*, a substance chemically allied to starch, and owing to the fact that many plant-cells develop hard, woody skeleton-cells formed of *lignin* or a thickened cuticle, the higher plants are usually much more stiff and rigid than the higher animals. It is much harder to plunge a dagger into an oak than into a man. Both animals and plants contain an enormous percentage of water, and this is equally true of the higher animals. Even the Archbishop of Canterbury comprises 59 per cent. of water. When you once traverse the skin, the interior of the body of a Vertebrate is about as soft as a very weak jelly. It is, indeed, semifluid, and thus the enormous numbers of parasites which make their way into the bodies of animals are able to push their way through the various tissues and arrive at the site they seek. Many animals are encased in external skeletons which protect their interior. Examples of these are the shells of molluscs, the hard casings of insects and lobsters,

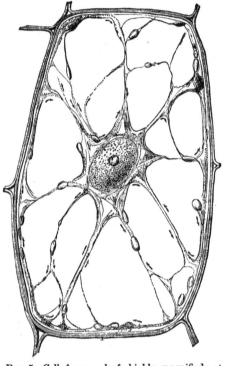

Fig. 5. Cell from a leaf, highly magnified: at the exterior is the cell-wall, which encloses the body: the latter is composed of a protoplasmic lining which surrounds the large watery vacuoles, while in the centre the nucleus is suspended by bridles of protoplasm. The arrows indicate the movements of the protoplasm during life. From Vines.

the scales of fishes, the armour of the armadillo, and the thick skin of the crocodile. But in man the skin is so soft that a female mosquito sinks her proboscis into it as easily as a butter-knife cuts into butter. The male mosquito is a feebler organism and its proboscis cannot penetrate the human epidermis.

Colonies

We have seen that when the *Amoeba* is dividing, the two halves are united by a bridge or waist, and for a time the organism is bi-cellular with two nuclei. But in higher Protozoa the bridge may not be broken, and if each of the new cells

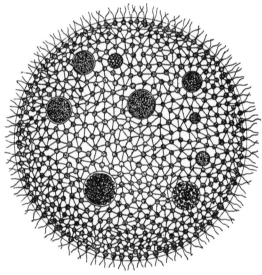

Fig. 6. *Volvox aureus* Ehrenb. A colony of the second order. (×210.)

divide again we have an organism of four similar cells, and if these four divide we have organisms of eight and then of sixteen cells, and so on. Thus we arrive at a multi-cellular organism which, if we regard each cell as an individual, forms a *colony*. Now there are three orders of colonies. The first consists of a collection of cells which have no intimate connexion with one another, whose protoplasms do not intermingle, and from which any cell can be separated without harming the colony: here each cell lives independently of the others. This is an example of a COLONY OF THE FIRST ORDER.

In the cell-colonies of the SECOND ORDER we find collections of cells which at first are very much alike, but they cannot live independently of the other cells that make up the colony. Their flagella, if they have flagella, move in unison. If one cell dies it must be replaced by another one. In colonies of the second order there is no differentiation of function. Each cell performs all the vital processes which are associated with life, though some cells may be set aside to form eggs and spermatozoa. But by far the greatest bulk of plants and animals consists of numbers of cells vitally related to one another, various functions being assigned to various sets of cells. Their tissues consist of a collection of cells each of which performs certain definite functions peculiar to that tissue. Thus we get muscle tissue, nerve tissue, digestive tissue, reproductive tissue, and so on. These form a COLONY OF THE THIRD ORDER.

INDIVIDUALS

These considerations raise the question as to what is an individual, and that is a problem which is very difficult to answer. We know quite well that the Prime Minister is an individual, and the policeman round the corner is an individual, so is a geranium, and so is a giraffe. But when we get down through the lower plants and animals which reproduce by budding the difficulty arises. On the whole, the cells that form a colony of the first order, entirely independent of each other, must be regarded as separate plants or animals and each cell as an individual. But when we reach the colony of the second order, where the cells interact and perform common functions, here the whole colony must be looked on as an individual of the same order as that of the higher plants and animals.

CHAPTER IV

FEEDING

THE FEEDING OF PLANTS AND ANIMALS—TYPES OF FEEDING—PARASITES—DIATOMS—BACTERIA

Feed me with the food that is needful for me.
Proverbs xxx. 8.

THE FEEDING OF PLANTS AND ANIMALS

ALL living organisms must have food. Protoplasm is perpetually wasting or burning away, and the loss must be made good or the plant or animal also wastes away, starves and in time dies. All sorts of substances must come from the outside, enter the body and be dissolved and distributed to the living cells. These substances contain the chemical elements which are found in protoplasm (C, H, O, N, S, P, Cl, K, Na, Mg, Ca, Fe). There is, however, no general, universal food which is suitable for all organisms, and, in fact, the food is almost as varied as are the organisms. The nature of the food enables us to separate plants from animals and fungi from green plants. Although apparently the protoplasm of plants and animals is the same, their food is different. Animals can and do consume solid food; plants in general cannot, and they must have food in a liquid or gaseous state. Animals cannot assimilate and use as food simple *binary* chemical compounds, compounds built up of only two elements. They can only extract their nitrogen from proteins and their carbon from proteins or from carbohydrates or from fats. That is to say, they must be supplied with the *ternary* compounds, which contain at least three different elements. Now these elaborate substances are only formed by and found in other animals or plants. Plants, on the other hand, can live on chemicals which contain but two different elements, water (H_2O), carbon dioxide (CO_2) and ammonia (NH_3). These chemicals exist in the air, in the water and in

FEEDING

the ground. Plants do not have to go about seeking what they may devour. The simple elements which form their food flow around them. Therefore plants are immobile, *sessile*, as it is sometimes termed. The food comes to them just as the roast geese in Heine's description of heaven perpetually flew up to the angels offering them tureens of delicious soup. That plants are, so to speak, bathed in their food explains the fact that they exhibit the greatest possible surface to the air above and to the soil below.

The food that is taken in by plants and animals is often compared to the coal which is the source of the energy of the steam engine. But we shall see that there is more in it than that. Like the coal in the engine, the food is burned up inside the body, and so produces the energy which is turned into work and heat, and thus enables us to walk and think and labour and breathe—in fact, it is the source of the power that keeps the plant or animal, as a machine, going. But food in a plant or animal does more than provide energy. It is used in repairing the wear and tear of the machine, and if the resemblance were complete the coal in a steam engine would be able to replace worn-out parts. This, of course, it cannot do.

For an animal to be in health its food must contain proteins, hydro-carbons, carbo-hydrates, salts, and water. Proteins are in the main supplied by flesh foods, the hydro-carbons by fats, and the carbo-hydrates by sugar or starch. The necessary salts of various kinds are contained in the above-mentioned foods.

We have said that plants can live on simple binary compounds. They can build up their protoplasm out of inorganic material only. But there is always an exception, and FUNGI (mushrooms, mildews, moulds), which are devoid of chlorophyll, have, like the animals, to obtain their energy and food by absorbing organic compounds. This is true, at any rate, as regards their carbon. Carbon that fungi take in must be taken up from organic matter although they can, like green plants, obtain their nitrogen from inorganic salts. They are thus in their feeding habits half-way between plants and animals, and this method of feeding is called *saprophytic*. They have to live on or near to organic food material. Thus you find

TYPES OF FEEDING

mushrooms and toadstools on a soil full of decomposing organic matter, and you find mildews and moulds living on jam or meat or cheese. But we have an exception to this exception. There are certain bacteria which, being devoid of chlorophyll, are fungi, and yet these can derive both their nitrogen and their carbon from ammonium carbonate, so that, like the green plants, they can live exclusively on inorganic foodstuffs.

Types of Feeding

There are, in short, four types of feeding:

I. HOLOPHYTIC. This is the method of green plants, whose food is inorganic, aqueous or gaseous.

II. SAPROPHYTIC. Characteristic of fungi, plants devoid of chlorophyll. Their food is organic and derived from dead organic matter or the product of living organisms and is absorbed in a liquid form.

III. HOLOZOIC. Peculiar to most animals. The food of these is organic and is absorbed in the solid form.

IV. PARASITIC. This is a mode of feeding common to such plants and animals as ingest organic food from some other plant or animal, the *host*, in or on which they live.

Plants have neither mouth nor stomach. They can absorb their food nearly all over their body. Hence it is to their interest to present as large an area as possible to the foodstuffs that always are flowing around them. As is well known, the average percentage composition of dry air is as follows:

	By volume	By weight
Oxygen	21·00	23·2
Nitrogen	78·06	75·5
Argon	0·94	1·3

In addition there is rather more than 0·03 per cent. of carbon dioxide by volume.

It is in the interests of the plant to present the largest surface to the air that it can. This is equally the case with the roots which permeate in all directions the moist soil, taking up water and the various salts dissolved in the water. It is often said that the roots of a tree present to the soil a superficial area comparable to that which the trunk, boughs and leaves present to the air.

Animals, on the contrary, have to seek their own food supply where they can find it. Thus they have a compact shape which facilitates movement, and they have a stomach and alimentary canal to digest and absorb the food when caught and eaten. Further, they often have to fight for their prey, which usually resents being caught and eaten. "Cet animal est très méchant, quand on l'attaque il se défend" is pretty well true of them all; thus they have developed organs for locomotion, claws and teeth and poison glands for defeating their prey, paws and hands for grasping it, mouth and stomach and alimentary canal for digesting it.

Parasites

There are, however, certain animals which do live surrounded by their food in solution. Parasites such as tapeworms, Cestoda, and the thorny-headed roundworms, Acanthocephala, which live in and on the nutritious juices of the alimentary canal of Vertebrates, have lost their mouth and intestine and absorb through their skins the more or less digested food of their host. The intestine of a Vertebrate is not an attractive abode, but even a tapeworm must have some place it can call a home.

We have seen that the great differences in the shape of the bodies of plants and animals are due to the different nature of their food. Plants can absorb their food in simple chemicals from the surrounding air, water and earth. Hence they remain stationary and do not move about. They have not to go and seek their food; the pleasure of the chase is denied them. Their food must be intolerably monotonous and extremely insipid—that which is derived from the air is for the most part just flat soda-water with the "sparkle" left out. Yet one hears no complaint. On the other hand, animals have to seek for their food wherever they can find it. As a rule, larger animals live on smaller animals or on plants and the higher animals live upon the lower animals or upon plants, but ultimately they are dependent on unicellular organisms.

Diatoms

In the long run marine organisms, including the fishes and even such mammals as whales and seals, depend for their food very largely on certain minute brown plants called DIATOMS, which play the greatest part in producing organic substances in the sea. They have been called the "pasture of the sea," and compared to the grass of our fields. Diatoms are found everywhere where there is water and where sunlight penetrates, at the surface of the sea and at the bottom of the shallow freshwater pools, in rivers and in lakes. They secrete skeletons of flint which, when they die in the ocean, drop to the bottom and cover some millions of square miles of the depths of the ocean. Their protoplasmic body with its nucleus is enclosed in a flinty shell shaped like a pill-box, and in it is found a *chloroplast*, or that unit of protoplasm which is set apart to contain the green colouring part of a plant, the *chlorophyll*. After a time of growth the body splits in two and each half of the pill-box then forms a new shell to replace the one that is separated off with the sister cell. As this process involves a diminution in size of the diatom, it would ultimately go out like Alice very nearly went out when nibbling cake in the "Wonderland." To avoid this humiliating experience, at certain times two diatoms come together and their protoplasmic contents leave the shells and fuse, and from this fusion emerges a diatom of typical size and form.

Their flinty skeletons are pitted or scored into wonderful symmetrical patterns, and so fine are these that they are used to test the powers of the highest lenses of our microscopes. Although diatoms contain but 10 per cent. of protein, 2·8 per cent. of fat, 22 per cent. of carbohydrate, and 65·2 per cent. of ash, they are so abundant that the higher life of the sea is mainly dependent on them as an ultimate source of food. They in their turn are, of course, dependent on the action of sunlight and cannot live below the level to which the rays of the sun penetrate. It has been shown that 1,200,000 diatoms of three species (and the number of species is enormous) exist beneath each square metre of surface over considerable areas of the North Sea. The incalculable number of diatoms that live near the surface of the sea is shown by the fact that their skeletons cover an area of about 10,880,000 square miles of the bottom of the ocean, principally in the southern waters. In one haul, which traversed two cubic metres of sea, there were no less than 3,173,000,000 diatoms taken; and yet it is believed that two-thirds of those present when the collection was taken escaped through the meshes of the fine net. Observations made in the Channel during the late winter months of sunshine on the number of diatoms enable us to some extent to foretell the success or otherwise of the mackerel fisheries of the later months of the year.

Besides the diatoms there are innumerable other unicellular plants and animals and an immeasurable number of minute larvae of animals floating about in the sea and in our rivers and lakes. Some of these are so minute that they readily slip through the meshes of the finest fabric man can produce. Fortunately Nature has come to the rescue, and the feeding organs of a certain sea-squirt, an ASCIDIAN, *Oikopleura* by name, net the very minute floating organisms which cannot be caught otherwise, though they may be separated in bulk by using a centrifuge. The floating and, for the most part, unicellular plants are collectively known as PHYTOPLANKTON, and the Phytoplankton of the sea, though composed of minute plants, is far vaster in quantity and bulk than all the *littoral* or shore seaweeds put together: the latter are relatively unimportant as a source of food for sea animals.

Diatoms are far more numerous in Arctic and Antarctic regions than in the tropical seas and with this is associated the enormous abundance and the colossal size of the circumpolar marine life. A certain jelly-fish, *Cyanea arctica*, is said to attain a diameter of thirty feet with tentacles 200–300 feet long. Also a certain cuttle-fish, *Loligo*, may become eighteen feet or more in length, whilst its allied forms from temperate seas are but some three inches long. Crustacea allied to our sandhopper attain in these polar regions the dimensions of lobsters.

BACTERIA

There is another group of lowly plants, microscopic in size, which are in many ways indispensable to the life of higher

BACTERIA

organisms. They take an almost equal part in our life and in our death, for whilst some are beneficent, many are malignant. In a later chapter we shall see what a large part they play in the soil, and we shall see how certain plants could not live without their aid. In shape they may be spherical, *Coccus*; or shaped like a rod, *Bacillus*; or spiral, *Vibrio* or *Spirillum*. Sometimes they are grouped in packets, as in *Sarcina*. These obscure plants are sometimes motile, being provided with cilia which vary in number and position. Although there are traces of nuclear matter, they have no definite nucleus. They multiply by dividing in two, and at an appalling rate. The ordinary hay bacillus, *B. subtilis*, divides about three times an hour, so that in the course of eight hours a single specimen will have produced over sixteen million offspring. They are very susceptible to light and are readily destroyed by the blue-violet rays of the spectrum. This, of course, has a vast influence on public health, as many of the bacilli are disease-causing germs. Their most important function is that, like other parasites or saprophytes, they break down complex organic compounds into simple ones; and although their methods of doing this vary widely, the final products are chiefly water and carbon dioxide. In the course of breaking down these complex organic compounds many by-products are produced, some of which are extremely harmful, and even fatal, to man and other animals.

They live in all sorts of queer places. One genus which oxidizes iron sometimes chokes the water-mains of large cities. Others will live in sulphur springs and separate the sulphur from the sulphuretted hydrogen which makes the waters of Harrogate, for instance, so unpalatable. They are the essential organisms of putrefaction and they play a large part in commercial products. The flavour of tobacco, butter and cheese is dependent on them, and they take a large part in the preparation of flax and hemp and indigo. Though we cannot get on without them, in a large number of cases we cannot get on with them. They are practically omnipresent, and we breathe them with every breath of air. They infest all the waters of the world, and are found throughout the superficial layers in the soil.

CHAPTER V

CHLOROPHYLL

CHLOROPHYLL—THE OXYGEN AND CARBON DIOXIDE CYCLE—MINERAL OIL AND NATURAL GAS—HAEMOGLOBIN

> La matière verte des végétaux....Nous proposons de lui donner le nom de chlorophyle.
> PELLETIER ET CAVENTOU (1818).

> When I want to know why a leaf is green, they tell me it is coloured by "chlorophyll," which at first sounds very instructive; but if they would only say plainly that a leaf is coloured green by a thing which is called "green leaf," we should see more precisely how far we had got.
> RUSKIN, *Queen of the Air* (1869).

CHLOROPHYLL

A GREAT annual miracle occurs each spring-time in the temperate regions of our world. The "boyhood of the year" grows up. Amongst flowering plants, the snowdrops and crocuses, the anemones and daffodils take the lead; amongst the trees the elder and the horse-chestnut usually begin it.

Tennyson tells us—

> The tender Ash
> Delays to clothe herself when all the woods are green,

In the autumn she is loath to lose her leaves, and they linger on when other trees are stripped of their green mantle, though the leaves rarely hang long enough to turn yellow. In the course of a few weeks most of our trees and shrubs and plants have clothed, as it were, the earth with a green carpet; green, the colour of hope.

What is it that causes this wonderful verdure? It is a substance called *Chlorophyll*, the most wonderful substance in our world. A world without chlorophyll would be a world without the higher forms of life, and in such a world no life, save perhaps that of the lowest bacteria, could possibly endure. In fact, without this remarkable pigment the living world as at present constituted could not exist.

CHLOROPHYLL

If you examine the cells of the green leaf you will find in their protoplasm small spherical discs, known as *chloroplasts*; and the chlorophyll is confined to these chloroplasts and, as a rule, is not diffused through the protoplasm, though there are exceptions. In *Spirogyra*, a fresh-water alga, the chloroplast is in the form of a spirally wound band. Chlorophyll consists of a mixture of colouring matters, the two most dominant of which are green and the others yellow. It is not soluble in water, or it would be dissolved out in the sap; but it is soluble in alcohol and in certain other fluids.

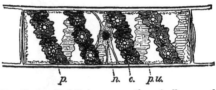

FIG. 7. A cell of *Spirogyra*. *c*, the spirally wound chloroplast; *p.u.*, the protoplasm lining the cell; *n*, the nucleus suspended by protoplasmic ropes; *p*, numerous small starch grains. (Darwin's *Elements of Botany*.)

Now the most wonderful thing about chlorophyll is that in the sunlight it can build up the plant's simple food, carbon dioxide, CO_2, and water, H_2O, into sugar and starch, setting free a certain amount of oxygen, which leaves the plant as an excretion.

The Oxygen and Carbon Dioxide Cycle

The first stage of the process is thought to be a chemical compound called *formaldehyde*, and a method has quite recently been discovered by which formaldehyde can be photosynthetically produced outside the plant's body. But as yet formaldehyde has not been found in the cells of the leaf. If it occurs it is only momentarily and is probably at once converted into sugar (glucose). The soluble carbo-hydrates (sugar, etc.) with certain simple salts build up a series of elementary nitrogenous compounds known as *amino-acids* and these by further synthesis lead to *proteins*. Amino-acids can be separated out from proteins and can be artificially made in the laboratory. By subjecting carbon dioxide and water to rays of light of very short wave-length, far beyond the violet end of the spectrum, formaldehyde has been produced. As these rays of short length are not found in sunlight, the process that goes on in the laboratory is not the same that

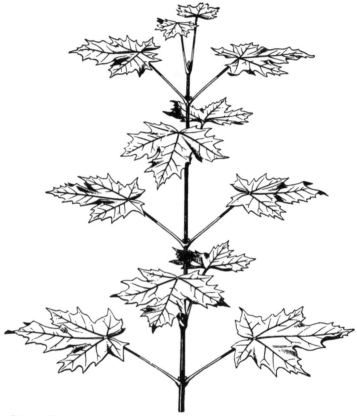

FIG. 8. Erect shoot of Norway Maple, *Acer platanoides*, viewed from the side and from above; showing the horizontal display of the leaves to catch the most sunlight. Kerner.

goes on in the plant—if it goes on. It was then further discovered that certain pigments such as methyl orange, when added to the mixture of water and carbon dioxide, acted like

PHOTO-SYNTHESIS

the chlorophyll in a plant, and caused the synthesis of formaldehyde to take place in ordinary sunlight. The formation of formaldehyde appears to be dependent upon the finely dispersed iron within the chloroplast, while the function of the chlorophyll itself is probably to convert the formaldehyde into sugars. If the formation of this chemical formaldehyde and the amino-acids be the first step by which the living plant builds up its starches and sugars, these experiments at least hint at a possibility of preparing food material in the laboratory, food which has hitherto been only and exclusively produced by plants. The process of combining the simpler substances into more complex compounds under the influence of light is called *Photo-synthesis*.

Photo-synthesis does not require direct sunlight, and it has been shown that it is carried on most successfully in land-plants when the intensity of light is about one-fourth of the intensity of full sunlight. And the same seems to be true of marine diatoms. The very brilliant light may actually damage the chlorophyll and other living tissues of the plant, just as the actinic rays of sunlight have a powerful effect on the human body. Some plants are able to protect themselves against this great intensity by manufacturing pigment in the epidermis, just as the human skin may protect itself by turning brown. A slight shade seems to be the best light for photosynthesis. It is curious to reflect that the sun never sees the shade, and yet it is a very obvious fact. From the point of view of the sun the shade is like Mr Chevy Slyme in *Martin Chuzzlewit*, "always waiting round the corner":

> Never in its life has the sun seen shade,
>   Never in its life seen a shadow where it falls:
> There, always there, in the sun-swept glade,
>   It lurks below the leaf; behind bodies, under walls,
> Creeps, clings, hides. Be it millions, be it one—
>   The sun sees no shadow, and no shadow sees the sun.
>
> LAURENCE HOUSMAN.

The great difference in the shape of the bodies of plants and of animals is largely due to the difference in their food. We have seen that plants can absorb their food in simple chemicals from the air and the earth. Hence they are as a

32 CHLOROPHYLL

rule rooted and immobile. As the prophet Isaiah said, "their strength is to sit still." On the other hand, animals have to seek for their food wherever they can find it. They have to move about to seize it, and hence their bodies are compact. They have to retain the food whilst it is being digested and built up into protoplasm; therefore with the exception of certain parasites they have to have an "interior," a stomach and an intestine. If they be carnivorous they have to develop organs for killing and holding their prey.

The actual feeding of the plant by the air is conducted as follows. The atmosphere contains about 0·03 to 0·04 per

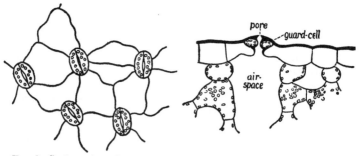

Fig. 9. Surface view of piece of lower epidermis of London Pride, showing five stomata, and a section through a stoma, showing the guard-cells and pore, and the large air-space immediately below the pore.

cent. in volume of carbon dioxide. The percentage of carbon dioxide in air is thought by some to be regulated not so much by the amount given off from the green plant as by the amount of the dissociation of the normal and acid carbonates of the sea. The carbon dioxide finds its way into the tissues of the leaf through the *stomata* or small pores which lead into the intercellular spaces of a leaf, for a leaf is traversed by small channels. In this way the carbon dioxide passes into the chlorophyll-containing cells. At the same time water is absorbed by the roots from the moist soil and passes up to the leaves with the ascent of the sap. Then, owing to certain wonderful powers that it possesses, chlorophyll is able by the energy which it obtains from the sunlight to build up the

FORMATION OF STARCH AND SUGAR

carbon dioxide and the water into sugar. Thus a carbohydrate is formed, and thus is the plant's food built up. The sugar in solution slowly passes down the veins of the leaf and leaf-stalk into the main part of the plant. During the day time more sugar is usually formed than can be transported by this slow process and the result is that the sap of the leaf cells becomes more concentrated—sweeter. An indefinite increase in the concentration would clog the mechanism, and it commonly occurs that where the sap contains more than a certain percentage (usually about 0·5 per cent.) of sugar, all in excess is transformed into insoluble starch. The starch exists in small granules in the chloroplasts, and if the plant be removed to darkness it will gradually disappear, because the transport of sugar is going on the whole time and as soon as the concentration of sugar in the sap drops below the critical figure, starch is turned back to sugar once more.

The sugar may be immediately used up in a growing plant, or it may be stored away as a food reserve for the future. For instance, carrots and turnips store up grape sugar, and sugar-cane and beet-root, cane sugar.

All the time this building-up process is going on—and it takes place only during daylight—the living matter of the plant is absorbing some oxygen and giving out some carbon dioxide; and this it does during its whole life, whether in sunlight or in shade. This latter process is called *respiration*, and it is common both to plants and animals. But the plant is breathing out what it wants back as food, namely carbon dioxide, and breathing in what it excretes as the waste products of the food, namely oxygen.

As Sir E. Ray Lankester has written, this action of chlorophyll is "the critical step in the interaction of chemical elements on the earth's surface, by which life is at present determined. Were there no assimilation of carbon from carbonic acid to form sugar or starch—by green plants, the whole fabric of the living world would tumble to the ground—in truth become mineralized. All living matter breaks down, within a short space of hours or days, to the resting or mineral condition of carbonic acid and ammonia (or nitrates). Were the building-up process, the raising to higher potentiality, not incessantly performed by green plants—a power which chlorophyll and chlorophyll alone confers on them—all carbon must pass from the reach of the organic world and living matter come to an abrupt end." Thus it comes about that animals are ultimately and absolutely dependent for their complex food on the green colouring matter of plants. Carnivorous animals are in the end dependent on herbivorous (plant-eating) animals, and thus both are dependent on the food which the plant has built up, not only for the increase in their bulk but for the energy which enables them to live and move and have their being. It is chlorophyll which "makes the world go round." If anything went wrong with this green substance—and it might—the whole of living creation would soon cease to be and the world become as dead as the moon is.

The fact that a plant is bathed in its food, as a baby would be if it were immersed in a bath of milk or beef tea, renders it important that it should expose as much of its surface as possible to the circumambient nutriment. Hence, leaves are numerous and flattened. It is not easy to find data as to the number of leaves on a plant; but many years ago there stood in Garden Street, near the Common at Cambridge, Massachusetts, a great tree known as the Washington elm, for it was under this tree that Washington in 1775 took command of the American Army. The ravages of time and of the "leopard-moth" have reduced this tree to a stump, but at its prime it was calculated to bear some 7,000,000 leaves whose surfaces added together amounted to about five acres. In sunlight this vast area was building up foodstuffs out of carbon dioxide and water.

Mineral Oil and Natural Gas

But there is a further consideration which shows the immense importance of this green colouring-stuff. It is well known that coal is the mineralized product of innumerable plants living in ancient geological times, and the same seems to be true of the other great stores of energy which are at the disposal of mankind.

MINERAL OIL AND NATURAL GAS

The so-called mineral oil and natural gas, the most modern sources of light, heat and power, are now believed to be directly of organic origin. Several writers, including the famous Russian chemist Mendeléeff, have supposed that oil was formed by the action of water on metallic carbides existing in the heated interior of the earth, a process exactly analogous to the formation of acetylene from calcium carbide, but the balance of opinion now inclines strongly to the view that oil and gas are formed by the natural distillation of organic material entombed ages ago in the rocks. Whether this material was mainly animal or vegetable is still a matter of dispute. If vegetable, it must have been mainly seaweeds, since oil is usually found in rocks of marine origin and is nearly always associated with salt water. Probably in some instances both Foraminifera and Diatoms played an important part, but most oil is now generally believed to be formed from the soft parts of marine animals. Many large fossil Brachiopods found in the Carboniferous Limestone in Derbyshire and Yorkshire contain lumps of bitumen exactly like the residue left on distillation of oil, *i.e.* asphalt. The essential feature of oil formation is evidently *anaerobic* (without oxygen) decomposition of organic matter, the carbon and hydrogen uniting to form hydrocarbons, with elimination of oxygen and nitrogen; whether bacteria played any part in the process is still an open question. At the present time large quantities of oil are obtained in Scotland by distillation of shales rich in organic matter, especially carbon, probably including material of both animal and vegetable origin, as in modern estuarine and marine muds. A product almost exactly like natural oil can also be obtained by distillation of coal, which is certainly mainly vegetable, but this is a costly and unpractical procedure.

With the exception of a small amount of energy which we can obtain from the tides, the rivers and the waterfalls, or by the pressure of the wind on windmills, the whole energy which man uses in his manifold operations depends on coal or oil or mineral gas. Without these resources of energy man would soon become as "the beasts that perish," and it is all due to chlorophyll, surely one of the most wonderful substances in the whole creation.

HAEMOGLOBIN

In this animal kingdom there is a pigment which may contest with Chlorophyll the title of being the most interesting material known. That pigment is *haemoglobin*. It is a much more complex substance than chlorophyll, being a compound of the pigment proper—haematin—with a protein globin.

To this pigment the higher animal creation, as we know it, owes its existence. The life of the mammal, as compared with that of the jelly-fish, is one of intensive oxidation. Intensive oxidation implies extensive oxygen supply and extensive oxygen supply can only be furnished to tissues far removed from the atmosphere by an efficient system of transport. Such a system is furnished by haemoglobin and there is no other material capable of playing the same rôle with equal success. Owing to its presence a given quantity of blood can take up seventy times as much oxygen as would otherwise be the case, and, what is almost more remarkable, the properties of haemoglobin and those of the human body are so adjusted that the oxygen is given up with the same facility in the capillaries of the tissue as it is acquired in those of the lung. The property of uniting with oxygen belongs to the haematin, the property of giving it up *in vacuo* is conferred upon the haematin by its union with globin, as also is its great solubility.

It is not a little remarkable that haematin and chlorophyll have much in common from the chemical point of view.

CHAPTER VI

THE NITROGEN CYCLE

SOURCES OF NITROGEN—THE PROTOPLASMIC CYCLE

> If in our withered leaves you see
> Hint of your own mortality:—
> Think how, when they have turned to earth,
> New loveliness from their rich worth
> Shall spring to greet the light; then see
> Death as the keeper of eternity,
> And dying Life's perpetual re-birth!   W. L.

SOURCES OF NITROGEN

WE have seen how a plant builds up its sugar and starch from the carbon dioxide of the air and the water absorbed by the roots from the earth; but to build up their proteins plants also require nitrogen. The atmosphere contains some 78 per cent. of this colourless, inert gas, which combines with difficulty with other elements.

There are several sources of nitrogen compounds in Nature, both inorganic and organic:

(i) Occasionally during a thunderstorm electricity causes the atmospheric nitrogen to combine with oxygen and to form nitrous or nitric acids which are brought down to the earth by the rain.

(ii) Perhaps the *Aurora Borealis* is an important source of nitrogen compounds. Astronomers have recently shown that there are always some auroral effects in progress, so perhaps we have here a continuous action similar to the synthesis of nitrogen compounds in a laboratory by means of the electric "silent discharge."

(iii) But a far greater supply of nitrogen taken up by plants and ultimately by animals is organic and is due to innumerable microscopic organisms called Bacteria. Certain of these organisms living in the soil and in the water, using sugars as a source of their carbonaceous food, are yet able to fix the nitrogen of the atmosphere. One of them, *Clostridium*, flourishes best in the absence of oxygen; that is to say, it is *anaerobic*. There is another one also occurring in sea water known as *Azotobacter*, which in the presence of oxygen is capable of fixing free nitrogen.

(iv) There is yet a further method by which the same gas can be built up into the food of the green plant. If one examines the roots of clovers or vetches or peas or beans or any other plant belonging to the family LEGUMINOSAE one will find certain swellings or nodules looking like tumours on the finer roots. Within these nodules are a number of bacteria of the genus *Pseudomonas*. Normally these bacteria live in the soil, but when they come across the thin wall of a root-hair of a suitable plant they press their way through it and then, quickly multiplying, they infect the tissues of the root and cause the formation of nodules. As they propagate, the cells of the nodules become swollen. Many millions of such bacteria may be found in the roots of one plant. The bacteria use the ready-made sugar which the plant has built up as nourishment, and in return they take up from the air, which is circulating in the interstices between the cells of the root, free nitrogen. After a time the green plant re-asserts itself and digests the bacteria, using up the nitrogen which the latter have fixed. On the death of the plant a few survivors remain and serve to re-infect the soil and other roots, and the nitrates which have been built up in the nodules help to fertilize the soil. The use of such plants as clovers, peas, and beans as rotation crops has been known for centuries. They greatly enrich the soil.

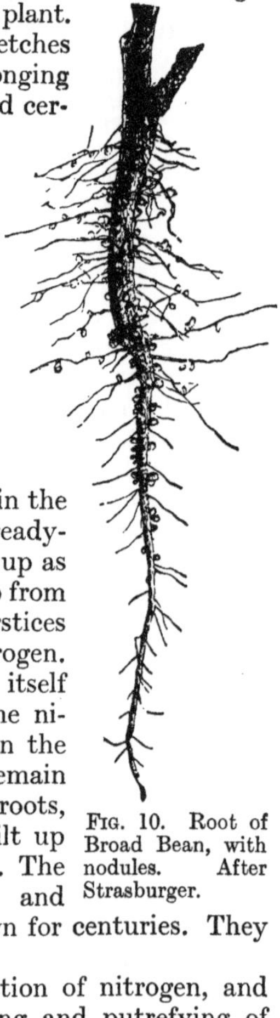

FIG. 10. Root of Broad Bean, with nodules. After Strasburger.

(v) Still another source of the fixation of nitrogen, and a most important one, is the decaying and putrefying of plants and animals and their refuse. The nitrogen in this

NITROSOMONAS AND NITROBACTER

case is already fixed and must at some previous period have been derived from atmospheric nitrogen from one of the sources already mentioned (i–iv). Such material, of which there is a plentiful supply in fertile soils, is known as humus. Decaying matter contains nitrogen locked up, and normally it serves as food for bacteria and fungi and other saprophytic organisms which contain no chlorophyll. Some of these are able to convert the dead nitrogenous matter into forms which are available for the plant to take up. If certain putrefactive organisms are absent or external conditions prevent their activity, decay does not take place and one sometimes finds dried-up mummies of animals who have escaped what is believed to be the universal fate. As animals decay ammonia is set free; this may escape into the air and be washed down to the soil by the rain. Here the ammonia forms ammonium carbonate. Now there is all over the world a bacterium known as *Nitrosomonas*, which is capable of converting ammonia into nitrous acid. Nitrous acid can combine with various salts, such as soda, or potash, or lime, to form nitrites, and there is still another bacterium equally widely distributed, known as *Nitrobacter*, which converts the nitrites into nitrates, the commonest source of nitrogen for green plants.

Finally there is one group of bacteria which act in the reverse direction, taking in nitrogenous compounds and liberating the free element. These "de-nitrifying" forms are prevalent in badly aerated or ill-drained soils and probably account in part for their relative sterility.

For purposes of illustration we may compare the organic world with a business in which nitrogen is substituted for money. When the balance sheet is made out, we have on the credit side of the account firstly the nitrogen derived from decaying organic matter by bacterial action, and secondly atmospheric nitrogen rendered available by bacterial and electrical action. On the expenditure side there are two items: one is a very large one, being the constant using up of nitrogen compounds by both animals and plants; the other is the loss due to the denitrifying organisms.

In Nature, as in business, the accounts must be made to balance, since if the expenditure is greater than the credit then life cannot continue owing to lack of food. If the balance

THE NITROGEN CYCLE

is on the credit side then all is well, but this rarely, if ever, happens in Nature, since in practice nitrogen is being restored to the air as fast as it is being taken out.

In the soil the "plant-available" inorganic nitrogen compounds are nitrates or ammonia—the nitrates form by far the larger source of nitrogen compounds on which the plants depend. Nitrates are soluble in water, and are thus taken up by the roots; in combination with the carbohydrates they are built up by some wonderful process not yet fully understood into the proteins, which are absolutely essential to the life of the higher plants and animals. At present there is a great gap in our knowledge subsequent to the passage of the nitrates from the soil into the plant and the appearance of the complex proteins which form the basis of protoplasm. But, as with the carbohydrates, we see there is a certain circulation or rhythm. The decay of plants and animals restores nitrogen to the air, and so does the action of the denitrifying bacteria. Plants and bacteria which fix nitrogen bring it back into the circle as it is built up into nitrates and ultimately into proteins. Animals get more nitrogen than they want, and the excess is excreted from the body in the form of urea, ammonium salts and uric acid. Whatever excess of nitrogen the plant possesses, and it is not much, is released from the body in the parts which die off at the approach of winter. The decay of both animals and plants is essential to the renewal of life in the world. But for the death and decay of living organisms, accompanied as it is by the breaking up of their complex proteins, life would cease. Were it not for the simple substances which green plants require as food, which are ultimately built up into the organic substances that alone can be digested by animals, the wheel of life could never revolve, and rhythm, save of the blindest of physical forces, would be quite unknown.

To every thing there is a season, and a time to every purpose under the heaven: a time to be born, and a time to die; a time to plant, and a time to pluck up that which is planted;...a time to break down, and a time to build up.

Ecclesiastes iii. 1–3.

The Protoplasm Cycle

Living protoplasm is ultimately built up by living organisms from certain substances which are found free in their environment. The most important of these substances are water, carbon dioxide and nitrogen; though small quantities of inorganic substances containing phosphorus, sulphur, sodium, potassium, calcium, iron, etc. are also required, as mentioned in Chapter II.

All living things absorb water directly from their environment, and they continually return water back to it as water-

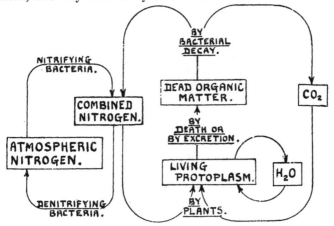

vapour or as liquid water. There is thus a continual interchange of water between living protoplasm and the environment, that is, a "water cycle."

Carbon dioxide and nitrogen have a more complicated history in protoplasm formation. Carbon dioxide is found free in the atmosphere and is utilized in building up protoplasm through the medium of the chlorophyll of plants; without chlorophyll carbon dioxide would no longer be available for protoplasm formation, and life—at any rate as we know it— would cease. When protoplasm dies, putrefactive bacteria set free the carbon dioxide again, thus completing the "carbon dioxide cycle."

THE NITROGEN CYCLE

Nitrogen, like carbon dioxide, is found free in the atmosphere, but it is in the elemental condition, and this is chemically extremely inactive. As we have stated, in the "nitrogen cycle" inactive atmospheric nitrogen is converted into combined nitrogen through the action of "nitrifying" bacteria. Others, the "denitrifying" bacteria, break down combined nitrogen back into the inert elemental form. Combined nitrogen can be transformed into protoplasm by plants.

The object of the diagram on page 41 is to show that these cycles are not to be considered as separate but are all part of one big "protoplasm cycle." The water cycle, the carbon dioxide cycle, and the nitrogen cycle are all mutually dependent; if one ceased, protoplasm could no longer be formed and the others would cease also.

Water is always available for living protoplasm, carbon dioxide is available through the mediation of plant life, while nitrogen is only available through the action of nitrogen-fixing bacteria and subsequently that of plants.

One important point is that nitrifying bacteria and plants are absolutely essential to the protoplasm cycle, whereas animals are only incidental; we can consider the whole of animal life merely as one of the ways in which living protoplasm is converted into dead organic matter.

CHAPTER VII

THE SOIL AND THE SAP

COMPOSITION OF THE SOIL—LIFE IN THE SOIL—THE SAP

> The thirsty earth soaks up the rain,
> And drinks and gapes for drink again;
> The plants suck in the earth, and are
> With constant drinking fresh and fair;
> The sea itself, which one would think
> Should have but little need of drink,
> Drinks twice ten thousand rivers up,
> So fill'd that they o'erflow the cup.
> The busy Sun (and one would guess
> By 's drunken fiery face no less)
> Drinks up the sea, and when h'as done,
> The Moon and Stars drink up the Sun:
> They drink and dance by their own light,
> They drink and revel all the night.
>   ABRAHAM COWLEY.

COMPOSITION OF THE SOIL. LEAF-MOULD

MUCH of the food of plants comes from the soil. Some soils are homogeneous, that is to say, they consist of a series of particles of the same or nearly the same chemical composition.

The sands of sea-beaches and of deserts and the coarse silts of rivers and fresh-water lakes are for the most part composed of nearly pure silica, but the finer silts and the clays contain a larger proportion of muddy material, which is essentially aluminium silicate, mixed with a variable amount of other substances, especially iron oxides, in an extremely fine state of division. The coarser-grained sands absorb water and retain it between their constituent particles, but more is absorbed by the finer silts and clays, which swell up when water is added and contract when it is removed. When moist these clays are sticky and can be easily moulded. Sand, on the other hand, has a lesser power of absorbing and retaining water, dries up quickly, and does not become sticky or form hard nodules or clods. Neither afford good land for plant growing. The chalk and the sedimentary limestones which are built up from the shells of minute marine organisms play a considerable part in the composition of fertile soils.

When we come to examine an ordinary garden soil we shall find that silicates and broken-up limestone, chalk and other rock-fragments, with certain phosphates, are mixed up with a great amount of organic matter, a rich dark-coloured material called *humus*. In course of time this is apt to spread over our chalk hills and our clay and sandy formations. The ideal soil is not composed of sand or clay or humus alone, but contains a proper proportion of all three: the sand to ensure porosity and a proper circulation of water, the clay to lend firmness and to prevent too rapid evaporation, and the humus to provide plant-food rich in nitrogen. The average grains making up a soil are about two and a half times as heavy as an equal volume of water, and the weight of a cubic foot of soil varies from about 80 to 105 lbs., the sandy soils being the heavier and those rich in humus the lighter. In ordinary soils from one-third to one-half of the volume is *pore-space*, which may be occupied by air or water according to circumstances, depending largely on rain-fall and the effectiveness of the drainage. The total surface area of the particles in one cubic foot of a light loam is estimated at about an acre. Each granule or particle of soil being surrounded by a pore-space, excepting at those points at which it is in contact with other particles, it follows that the air spaces, however contorted in size and direction, must form continuous tubes or passages which traverse the soil in all directions.

The organic matter has two sources. One is decayed plants and animals, which rot away and disintegrate in the soil. A familiar form of this is leaf-mould with its "moist rich smell of the rotten leaves," as Tennyson notes. That which has been built up from the soil is returned to it as decaying matter, "dust to dust." This decay is associated with a certain amount of heat, as one sees in a fermenting manure heap; and, in fact, as much heat is given off as would be given off if the decaying material were burned, but of course the process is much slower. It is this decaying organic matter that helps to make life possible in the soil.

BACTERIA AND PROTOZOA IN THE SOIL

LIFE IN THE SOIL

And here we come to our second source of organic matter. Garden soil is as teeming with life as is a great city. It is recorded that Prince Bismarck once said to Lady Randolph Churchill: "Have you ever sat on the grass and examined it closely? There is enough life in one square yard to appal you."

It has always seemed to me a strange thing for the Prince to have said. To begin with, throughout his long life he had shown but an imperfect sympathy with the lower members of the Animal Kingdom, and then, again, he was a man not easily appalled; but the saying is perfectly true. There are millions of bacteria performing various functions. There are millions of unicellular animals of an amoeboid and flagellate nature creeping in and out the interstices of the soil. As a result of 365 counts, made on consecutive days, of the bacteria and of six different species of protozoa in a natural field soil, it became clear that from day to day the numbers varied greatly, and that this variation had no connexion with the weather. Fortnightly averages showed certain seasonal changes, and the numbers of both bacteria and protozoa are greatest towards the end of November and smallest during February. The seasonal fluctuations resemble those of many aquatic organisms and are independent of rain-fall and temperature. There is a noticeable relationship between the numbers of certain *Amoebae* and of the bacteria. More *Amoebae* mean fewer bacteria. The former eat the latter.

It is easier to kill *Amoebae* than to kill bacteria. By an accident at Rothamsted it was discovered that certain soil heated to 98 degrees oxidizes more rapidly than normal soil and in it bacteria were increasing at a more rapid rate than usual. At the same time nitrates and ammonia in the soil increased very much more rapidly. There is evidently some living organism in the soil which consumes bacteria and is destroyed at 98 degrees Fahrenheit while the bacteria persist. By using such partially sterilized soil it is claimed that a crop of tomatoes which normally produced 32 to 35 tons per acre could be increased to 80 or 90 tons.

Minute round-worms are creeping and wriggling through the soil. Some of these spend their whole life there, others only their larval existence, whilst others return to the soil only when mature. A typical example of this is the round-worm which causes disease in adult grouse. It leaves the alimentary canal of the bird with the droppings and produces its young on the soil. The larvae make their way to the heather, and slowly wriggle up to the heather-buds which form the favourite food of the mature grouse. With this food they pass into the alimentary canal of the game bird, become adult, and set up fresh trouble in the bird's intestine. It is difficult for the layman to grasp what is going on in and on the soil and on the plants which it supports. Suppose we could by means of a gigantic lens magnify a square yard of a grouse moor one hundred times. The heather plants would be as tall as lofty elms, their flowers as big as cabbages, the grouse would be about six or seven times the size of Rostand's "Chantecler" at the Porte St Martin Theatre. Creeping and wriggling up the stem and over the leaves and gradually yet surely making their way towards the flowers would be seen hundreds and thousands of silvery-white worms about the size of young earthworms. Lying on the leaves and on the plant generally would be seen thousands of spherical bodies the size of grains of wheat, the cysts of the *Coccidium*—a protozoan parasite which destroys the young grouse chick—and on the ground and on the plants, as large as peas, would be seen the grouse tape-worm eggs patiently awaiting the advent of their second host. It is perhaps a picture which will not appeal to all, but yet it represents what, unseen and unsuspected, is always taking place on a grouse moor.

Then there are innumerable insects, chiefly in the caterpillar stage, which crawl through soft earth and do infinite harm to trees, plants and crops by living on or nibbling their roots. One of them, the *Cicada*, spends seventeen years as a larval form in the earth in the southern part of the United States. Many millipedes also live crawling in the soil. One of them, known as the wire-worm[1], does considerable damage. Again

---

[1] Another injurious animal, also unfortunately called a "wire-worm" is the larva of a beetle, *Elater lineatus*.

ANIMALS IN THE SOIL

we have the earthworm, helpful rather than harmful, which is perpetually wriggling its way through the soil, eating its way through it and extracting what organic matter it can get from the swallowed earth; and passing the indigestible remains in the form of worm castings on to the surface of the land. Darwin has estimated that these dried-up casts will raise the level of an acre of ground in England by one-tenth of an inch each year.

The soil, as we have seen, is simply teeming with life, while manured soil teems more. An investigation carried on at Rothamsted on two plots of ground, one of which had received no manure since the accession of Queen Victoria in 1837, and the other of which had received 14 tons of farmyard manure per acre, per annum, since 1843, showed that there were normally three times as many animals in the latter as there were in the former. In round numbers there were 15,100,000 animals per acre, of which 7,720,000 were insects, in the manured plot. The corresponding numbers in the unmanured acre were 4,950,000, of which 2,470,000 were insects. The greatest number, both of insects and other invertebrates, occurred in the upper three inches of the soil; but there were exceptions to this. Most of the injurious insects, such as the wire-worm, *Elater*, and the larvae of the daddy-long-legs, *Tipula*, were not affected materially by the manurial treatment of the plot. The invertebrata concerned consisted of larvae and imagos of all the principal groups of insects, many species of centipedes and millipedes and a certain number of spiders and of mites, many worms and round-worms and certain terrestrial crustacea, such as the wood-louse; and a number of snails were also found.

There are many other burrowing animals, such as the marmot and the prairie dogs, the rabbit, the fox and the badger, but the most important of these in our country is the mole. This burrower displaces a very large quantity of earth. It works very hard, has a tremendous appetite, eats all manner of insect grubs, and will consume its own weight of earthworms in the course of twenty-four hours. It seems to work rather spasmodically, as the earth castings are ejected at periods of three hours.

48   THE SOIL AND THE SAP

Now the effect of all these animals pushing and shoving through the soil is to open up passages for the air to enter, and the soil becomes aerated, and when rain comes these passages further convey water to the deeper layers of the earth. We have said that the soil is as teeming with life as a great city, and as regards the unicellular inhabitants there is the same sort of rhythm. The fall and rise in their number is comparable to that which a great city experiences when the number of the population is decreased by a Bank Holiday, or in the richer and more aristocratic quarters by the fatal habit of "week-ending." So numerous are these micro-organisms that it has been stated that "a single salt-spoonful of soil will contain millions of organisms, some active, others in repose, and many in the form of spores."

If we could only see a mass of soil under a microscope, we should find it as porous as a sponge. From 30 to 50 per cent. is empty spaces—pore-spaces—filled with air or with water, and the water contains plant-food in solution.

The Sap

In a moist soil each particle which helps to compose it is enveloped in a film of water, and when there is plenty of water about that water accumulates in the spaces which exist in the porous earth. This water is taken up into the roots by the *root-hairs*. The process by which the thin watery fluid of the soil with certain salts in solution is drawn through the organic cell wall and the protoplasm lining of the root-hair and mingles with the denser cell sap, is called *osmosis*. When certain chemical substances in a more or less concentrated solution are separated by a semi-permeable membrane from water with salts, etc., in a more dilute solution, the latter liquid will pass through the membrane and mingle

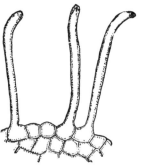

Fig. 11. A portion of a section through a young root, showing some of the superficial cells growing out into root-hairs. A thin layer of protoplasm (dotted) lines the cell-wall, and encloses the cell-sap.

ROOT-HAIRS

with the more concentrated fluid. The extent and the rapidity of this osmotic action depends (i) on the concentration of the stronger fluid, (ii) on the nature of the substances it holds in solution. Each root-hair is a microscopic tubular outgrowth from a single cell of the outermost layer—the *epidermis*—of the root. Inside it contains the cell sap, which is more concentrated than the water of the soil. The more concentrated contents of the root-hair are separated from the weaker water of the soil by a protoplasmic membrane and by the cellulose cell-wall, and the less concentrated fluid of the soil passes through the membrane into the cavity containing the more concentrated cell sap. These root-hairs greatly increase the surface area of the root which is capable of absorbing moisture. They grow out between the particles of the soil and are surrounded by the water, which in its turn surrounds the constituent particles. When the water contains a large amount of salts, as it does in some marshes, the osmosis is slowed down or even arrested; hence ordinary plants do not grow well in soils with much salt in them. The root-hairs seldom absorb the whole of the available water, for there is always a certain surface tension which favours the retention of water between the particles of the soil. A soil is rarely if ever wholly without moisture. In water-plants a certain amount of water is taken up by the leaves and stems.

Fig. 12. Seedlings of Mustard: *A*, with soil adhering to the root-hairs; *B*, with root-hairs free from soil. After Sachs.

Having once got the water with the dissolved salts (food substances) into the root-hairs and so into the roots, it has to be dispersed, and the question of the ascent of this fluid, now called *sap*, is a difficult one. It travels in minute cells called *tracheids* or in *vessels* which are formed from rows of superimposed cells whose adjacent, transverse walls have broken down, thus producing minute capillary ducts. So fine are they, 100μ or less in diameter, that a certain amount of sap may ascend by capillarity; but this would not carry it more than a few inches and would certainly not account for the arrival of the sap at the top of tall trees. We shall see later that the leaves are constantly giving off watery vapour, and this produces a certain "pull" on the sap in the tracheids. Now a column of water offers considerable resistance to being broken, as anyone can see who has studied the Chaine-Helice Patent Liquid Elevator. There is a tensile stress which keeps water in continuity; in a sense water is sticky and its continuity is not readily broken. If the sap of the upper end is passing away, evaporating, the water at the top of the vessel pulls up the water just below it, and the fine column of sap in the tracheids is not broken.

The tracheids and vessels are formed from dead cells whose protoplasm has disappeared. To prevent collapse their walls are thickened, and these thickenings may be spiral, ring-like (annular), ladder-like (scalariform) or like

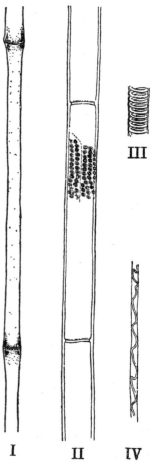

FIG. 13. From longitudinal sections of the stem of a Sunflower. I, part of a sieve-tube; starch grains are clustered near the sieve-plates. II, part of a pitted vessel, cut in half lengthways, showing the remains of two cross-walls; the pits are shown only in a small area of the wall. III, a small part of a spiral vessel. IV, part of a spiral vessel that was formed very early and has been greatly stretched during the growth in length of the stem: the spiral band has been pulled out and the whole vessel has collapsed. Magnified.

FIBRO-VASCULAR BUNDLES

net-work (reticulate). Vessels have the same function as tracheids, that is, they conduct water. In vessels a longitudinal channel is formed which has a greater diameter than that of the tracheids, *i.e.* 300 to 700μ. Owing to the cavity of the vessels being continuous from end to end the vessels are more efficient as water-conduits than the tracheids and through them most of the water passes.

FIG. 14. Leaf of Plane, *Platanus orientalis*, showing the fibres and vascular bundles. From Ettingshausen.

The organic food material is conveyed throughout the plant by the *sieve-tubes*. These are formed of rows of cells whose transverse walls are pierced by a number of minute holes as a sieve is. These are called *sieve-plates*, and through the holes in them the protoplasm of one cell is continuous with that of the next.

All the three conducting tissues, tracheids, vessels and sieve-tubes help to form the "fibro-vascular bundles" which traverse the plant from the root tip to the leaves.

THE SOIL AND THE SAP

The simpler plants which live in water have no special apparatus for the absorption or for the transpiration of the water with its dissolved salts. This passes through the general surface into the body of sea-weeds and other aquatic plants. Even the higher vascular plants such as water-lilies, which live in water, are to a great extent devoid of woody tissue, as the water has no need to be conveyed from the root to the leaves. It is in terrestrial plants where the absorption of water is almost confined to the root that the conducting cells leading upwards to the leaf are most highly developed.

The capillary action plays but a small part in the ascent of sap. A more important part is played by what is called *root pressure*. If you cut across a growing stem of a vigorous plant in the early part of the year sap may be seen exuding from the cut surface of the lower half, and this cannot be due to any "pull." It is attributed to pressure in the root. But here again root pressure does not account for more than a certain "lift" in the sap. It varies also from time to time, and is not constant in all plants. Great fir trees have a very small root pressure and yet their sap must be raised to very considerable heights, and root-pressure frequently seems to disappear in the height of summer when the demand for water by transpiration from the leaves is at its greatest.

Thus of the three forces which act in the ascent of sap, the only two appreciable ones are root pressure, or the "push" from behind, which is variable in amount and ceases at times, and does not at best raise very many ounces. Many more ounces are lifted by the "pull" from above, due to the viscous nature of a column of water. The whole thing recalls the poem:

> The Temple of Fame is open wide,
> Its Halls are always full.
> Many get through by the door marked "Push,"
> But more by the door marked "Pull."

The rate of ascent of sap varies under varying conditions. The force which is available for moving the sap is about 300 atmospheres. The resistance may be great or small according to the supplies of water in the soil and the height of the plant.

THE STRUCTURE OF A LEAF

Experiments on conifers growing under normal conditions of supply give the velocity at between 5 cm.–10 cm. per hour. If resistance is artificially removed, velocities of 3 m. per hour have been observed. High velocities reaching perhaps 50 cm. per hour probably occur temporarily in living plants, perhaps locally even more. There is plenty of evidence that the stream often comes to a standstill or is reversed in direction. Varying velocities and even reversals occur simultaneously in the same plant.

If we consider for a moment the structure of a leaf, we shall find that it consists in the first place of an outer single layer of cells, the *epidermis*. The external surface of each cell has become thickened and hardened and thus forms a cuticle, through which very little evaporation can take place. Beneath the upper layer of epidermal cells comes a second layer of brick-like cells, called *palisade cells*, arranged like a series of bricks on end. Beneath these again comes a *spongy tissue*, whose cells, which may be three or four layers thick, are separated from one another by spaces, sometimes fairly large, sometimes merely slits. Then comes the lower epidermal layer of cells. These spaces communicate with the external air by means of little valves called *stomata* which consist of two sausage-like cells, the *guard-cells*, lying face to face with a spindle-shaped aperture between them which opens as a result of the action of sunlight. Just within each stoma there

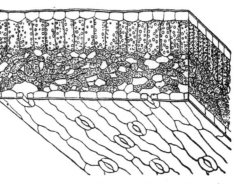

Fig. 15. Semi-diagrammatic view of a leaf in section, showing upper and lower epidermis, the latter with stomata. Between these two layers the mesophyll, the palisade tissue above, and the spongy tissue below. Two stomata in section show how the intercellular spaces communicate with the external air.

is an unusually large space. Now water that is coming up in the vessels is conveyed along the *veins* or *ribs* of the leaves, which gradually get smaller and smaller and finer and finer. Ultimately by osmosis the water passes into the spongy tissue of the leaf and then into the walls lining the spaces between the cells of the spongy tissue. These spaces thus become extremely humid and damp, and if the surrounding atmosphere is not saturated the watery vapour passes out by diffusion through the mouths of the stomata into the surrounding air. This process of getting rid of the water which was in the first case taken up by the root-hairs is known as *transpiration*.

The number of stomata is astounding. They are for the most part placed on the under surface of the leaf, except in floating leaves such as the water-lilies, where they are found only on the upper surface. They are extremely minute, the spindle-shaped opening being in the field spurry, *Spergula arvensis*, 0·007 mm. in diameter, whereas in some kinds of lilies, *Amaryllis*, they are as wide as 0·02 mm. A typical sunflower-leaf may have as many as 13,000,000 of these stomata and the amount of fluid which passes through them is very considerable.

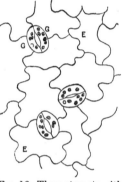

FIG. 16. Three stomata with surrounding epidermic cells (E); G, G, guard cells of a stoma.

It has been mentioned above that an elm tree may have 7,000,000 leaves whose surface areas when added together may amount to some five acres. A single birch tree with some 200,000 leaves will pour into the atmosphere 15½ gallons of water on an ordinary day, and on a very hot, dry day as much as 85 gallons. A sunflower with a leaf-area of 5616 square inches loses a pint and a half in 12 hours, and it has been estimated that a beech forest evaporates about 14,000 tons of water per acre during the summer months. An average acre of wheat will from start to finish give off during the life-time of the plant some 1000 tons of water. Certain Danish investigators have calculated that the

NUMBER OF LEAVES

119,000 leaves of an average-sized beech tree have a superficial leaf-area of 341 square yards. A fir tree with 39,680,000 leaves has a superficial leaf-area of 2180 square yards. A pine tree with 20,043,000 leaves has a superficial leaf-area of 940 square yards. The number of pine and fir needles is indeed amazing. A pine whose stem has a diameter of some 5 inches has millions of such leaves, and the fully grown firs and pines with a stem diameter of 20 inches have from 10 to 20 millions, whilst the largest specimens with a diameter of 30 to 35 inches will bear from 30 to 40 million needles. Some of the factors which are involved in transpiration are the number of leaves and their superficial area, the number of stomata, and the varying humidity of the surrounding air.

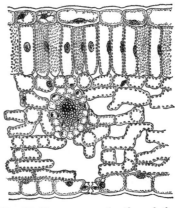

FIG. 17. Transverse section through the leaf of the Hellebore, showing, from above downwards, the upper epidermis, the palisade cells, the spongy tissue (in which a vascular bundle is seen), the lower epidermis, in which is shown a single stoma opening into a large intercellular space.

These figures give some slight idea of the enormous part that vegetation plays in the humidity of the atmosphere. It is to a large extent a fact that with the destruction of forests the humidity of the air falls and tends ultimately to disappear. Rain and fog and dew diminish, the soil becomes dried up and vegetation ceases to exist.

Apart from the tracheids, that bring the water with salts in solution from the roots to the leaves where by osmosis it enters the mesophyll cells, there is a system of vessels called sieve-tubes which take the elaborated food from the leaf to those parts of the plant which require it, either for their growth or to store it up as reserve food after it has been reconverted into starch. This acts as a reservoir or store of food for next year's growth.

Thus the carbohydrates, being soluble, diffuse down through the cells and possibly through the vessels, while the proteins, being insoluble products, must be conducted through tubes, the sieve-tubes. Typical examples of storehouses are the underground *tubers* which we call potatoes and the underground stem of the bracken fern.

The upper portions of a soft plant such as a stinging nettle are kept erect by the turgidity of their cells. The great majority of the cells are kept in a stretched condition by the pressure of the watery spaces inside their protoplasm, pressing it tightly against the cell wall. Should, however, the supply of water from the roots be cut off, the plant wilts, withers, droops and loses its normal and healthy shape. But in those plants which are not annual, which do not die down each autumn and reappear each spring, in such plants as trees and in the lower parts of most herbaceous plants, there is a strong skeletal element made up of cells which have lost their protoplasm, and whose cell walls have become immensely thickened so that there is hardly any cavity left in the cell. Some of these cells called fibres change the nature of their cell-wall into woody substance or *lignin*. Such fibres are elongated dead cells, interlocked so as to form a tough and hard skeletal tissue. When mechanically "shredded out" they are used in the making of paper and linen and certain coarse fabrics. The fibres with other skeletal elements make the bodies of woody plants much harder than the bodies of animals. If we except those animals which are protected by a hard outer skeleton, it is very much easier to hammer a nail into an animal than into a tree. The arrows which pierced the body of St Sebastian would not have penetrated the bark of a cedar.

CHAPTER VIII

FOOD

CHEMISTRY OF FOOD—VARIETY OF FOODSTUFFS AND FEEDING HABITS

"From whence, my friends, in a human point of view, do we derive the strength that is necessary to our limbs? Is it," says Chadband, glancing over the table, "from bread in various forms, from butter which is churned from the milk which is yielded unto us by the cow, from the eggs which are laid by the fowl, from ham, from tongue, from sausage, and from such like? It is. Then let us partake of the good things which are set before us!"

Bleak House. CHARLES DICKENS.

CHEMISTRY OF FOOD

BOTH plants and animals grow and at the same time their tissues waste away. To supply the material for growth and to replace matter which has wasted away and been excreted the eating of food is necessary. "Il faut manger pour vivre," as one of Molière's characters in *L'Avare* says. We have dealt with the food of plants in Chapters IV, V, and VI, and now we deal with the food of animals, which is far more varied than that of plants.

The food of animals, at any rate of the greater percentage, consists of certain chemical compounds:

1. Proteins.
2. Carbohydrates.
3. Fats.
4. Water.
5. Salts.
6. Vitamins.

Of these, the first three are organic in their origin, the fourth and fifth are inorganic. I am not quite sure whether the sixth and last are foods, but they are essential for normal nutrition. In any case, they come within the compass of one definition of food, "whatever sustains and supports growth."

One of the most perfect foods for mammals is milk, which contains all these compounds in proper proportions, and all young mammals live for a time on milk. Eggs are almost as perfect a food, but they do not contain quite enough carbohydrates for a mammal. Vegetable foods, such as rice, potato, wheat, contain an almost excessive amount of carbohydrates, whilst animal food, such as mutton, beef, pork, chicken, contains an excess of proteins for man, who has as a rule a mixed meal. "Loaves and fishes" with some green vegetable —for the vitamins—form a light and wholesome diet. For a human being the vegetable and animal foods should be mixed in suitable proportions.

The process of building up the dead food into the living protoplasm of the cells is known as *anabolism*, and the process of breaking down the protoplasm of the cells into the various waste products is known as *katabolism*. The two together, the building up and the breaking down, are often referred to as *metabolism*.

Variety of Foodstuffs and Feeding Habits

As we shall see, the food of animals is very varied. Many animals are fixed, or move about with difficulty. Amongst the molluscs the fresh-water mussel, common in our rivers, moves chiefly at night and rests by day, but it does not move very far, perhaps only fifteen feet in the course of twenty-four hours, or about a mile a year. The marine or edible mussel is anchored to rocks, etc., by stout threads, and the oyster by one valve of its shell, and in all these cases the food is brought into the neighbourhood of the mouth by the action of cilia; these set up a stream which carries inside the shell not only food but oxygen. Cilia also bring sea-water laden with nutritive matter into the bodies of sea-squirts (Ascidians) and sea-anemones—"those flowering stomachs," as George Meredith calls them, "which open to anything and speedily cast out what they cannot consume"—and into many other aquatic animals. The minute plants and animals which form the food of jelly-fish are entangled in mucus secreted along

MUD-EATING ANIMALS

certain radii of the upper side of the *umbrella*. By the action of cilia, this food is passed along over the edge to the lower surface of the umbrella and ultimately to the mouth.

A good many marine worms live on mud, and so do the sea-lilies, CRINOIDS, and the sea-cucumbers, HOLOTHURIANS. It is a peculiarity of all these animals that the wall of their alimentary tract, packed as it is with fragments of sand and other samples of the sea bottom, many of them with very sharp corners, is of extraordinary thinness, transparent and tightly stretched, so that one might expect that at any moment the intestine would rupture. But it does not seem to do so. This sand or mud is rich in organic matter, both living and dead, and this it is that supplies the nutriment to the mud-eaters. Marine worms mostly belong to the group POLYCHAETA, *i.e.* with many bristles. They often eat each other but they cannot digest the bristles, which work their way through the tissues to the outside, much as a needle often does when it has inadvertently been swallowed by a child.

A rough rule of feeding is that the bigger animals consume the smaller animals and plants. Many unicellular animals ingest and digest bacteria and unicellular algae. Certain of the most infectious bacteria can serve as food for protozoa such as the *Amoeba*, and apparently cause no disease in the consumer's body.

Amongst the more simple multi-cellular organisms pretty well any smaller animal or plant may serve as food, though many of the minute protozoa seek to protect themselves from being eaten by forming skeletons of lime or flint or by flinty spicules. As we get higher up in the scale we find animals are more apt to pick and choose what they eat. Earthworms consume great quantities of earth, and the surface soil is teeming with an abundant fauna and flora which affords them ample nourishment. The mineral fragments which cannot be digested are passed out on to the surface of the land as worm-castings. In an average soil there are some 150,000 earthworms to the acre, and the amount of earth extruded from earthworm burrows is considerable. Darwin weighed the amount of earth thrown up on two separate square yards in a single year and found it to be 6·75 lbs.

60 FOOD

and 8·387 lbs. respectively, which would amount annually to 14·58 or 18·12 tons per acre on an average. This lifting up of the earth has saved for archaeologists many objects of intense antiquarian interest, and helps to account for the fact that the earth is slowly rising, a fact which anyone who visits any ancient Saxon church with its floor below the present level of its surroundings can easily verify. Earth-worms are equally of value to the agriculturist, for they break up the soil, bring the rich sub-soil to the surface and open up channels for the air and water to penetrate.

Leeches for the most part live on blood, and they are not particular how they get it. They will attack cattle, birds, frogs and tadpoles, insects, snails and worms. Like the other blood-sucking creatures, they are capable of conveying disease from one animal to another, and certain fish-diseases are thought to be spread by their agency. Leeches were wonderfully abundant in our pools and streams, but with the drainage of the Fens and other waters they have for many years become scarce and now have to be imported. The same scarcity was very apparent to the poet Wordsworth, whose insatiate curiosity is recorded in the following lines in 1802—Wordsworth was always asking rather fatuous questions:

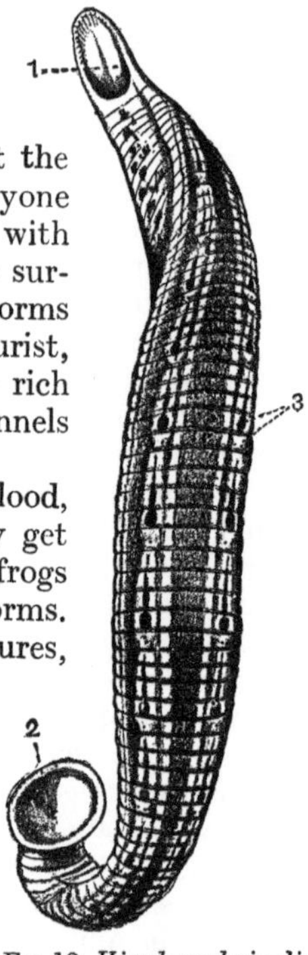

FIG. 18. *Hirudo medicinalis*, the medicinal leech. Life size. 1. Mouth. 2. Sucker. 3. Sensory organs.

> My question eagerly did I renew,
> "How is it that you live, and what is it you do?"
> He with a smile did then his words repeat:
> And said that, gathering leeches, far and wide
> He travelled; stirring thus about his feet
> The waters of the pools where they abide.
> "Once I could meet with them on every side;
> But they have dwindled long by slow decay;
> Yet still I persevere, and find them where I may.'

FOOD OF CRUSTACEA, CENTIPEDES, INSECTS

Leeches have many enemies—water rats, voles, the larvae of the *Dytiscus* beetle, the larvae of *Hydrophilus*, the *Nepa* or water-scorpion, the larvae of the dragon-fly and the adult *Dytiscus*—all feed upon them. Many birds also eat leeches; and it is recorded that at one artificial leech-farm, where there were 20,000 leeches, they were all eaten up in twenty-four hours by an invasion of ducks. Frogs and newts also devour them and they are not above eating their own brothers. *Aulostoma* will devour its own species as readily as it will an earthworm.

Crayfishes, which still lurk in the banks of some of our streams, have a very varied diet, living on dead animals or plants, worms, and the fresh-water alga, *Chara*, not a very edible vegetable, for it has a most unpleasant smell. So unpleasant is it that when grown in water in which mosquito larvae are living it proves fatal to them. Many of the smaller Crustacea feed on algae, protozoa and other aquatic organisms. Amongst these there is one large group of small forms called the COPEPODA which in their turn form the food of many marine and fresh-water fishes. They exist in enormous numbers and play a great rôle in the economy of the sea. A curious crustacean, the barnacle, is fixed on rocks or often to the bottoms of ships. The larva is free-swimming, but the adult lies on its back and literally kicks its food towards its mouth by the action of its six pairs of legs.

Some centipedes are carnivorous, but others are vegetable feeders, and one of the so-called wireworms, *Iulus terrestris*, often found in our country curled up under stones, does much damage by gnawing the tender roots of plants. Cockroaches will eat almost anything that man can eat and a great deal more, for instance other insects, such as the common silverfish *Lepisma*, paper, leather, paste, refuse of every kind, and even the dead bodies of their companions. They are also said to devour bed-bugs in spite of the unpleasant smell of the latter, and this habit is shared by a small black ant from Portugal. Dragon-flies, both in their larval and adult stage, will eat only living organisms. Like "living flashes of light" they hawk through the air, catching other insects in flight; whilst their larvae, more or less buried in the mud, devour worms, insect-larvae, small fish and tadpoles.

FOOD

Insect larvae are very voracious. The Daddy-long-legs in this stage will consume five to ten times its own body-weight daily. These larvae are further inveterate cannibals.

Amongst the animals that live on the skin of grouse are some very well-groomed, little, chestnut-coloured insects called "bird-lice." They live entirely on the smaller parts of the feathers, and as they take no fluid they enjoy a very arid diet.

Wasps are carnivorous, chewing up spiders, insects, and pieces of flesh torn from larger carcases. Bits of all these are carried by the worker wasp to the young larvae who, hanging upside-down in their papery cells, greedily devour them. Bumble-bees feed on pollen and a very watery honey, "weak but palatable" as Mr F. in *Little Dorrit* found the wines of France. The worker bee supplies the larva with a food which is secreted from its salivary glands. It is known as "pap" or "royal jelly" and has a white or yellowish jelly-like appearance. The worker fills the cell with this food and the larva not only eagerly laps it into its mouth but probably absorbs the pabulum, in which it floats, through its tender skin. On the fourth day after hatching the worker-larva is partially weaned and its food is now mixed with honey; twenty-four hours later the drones are completely weaned and are henceforth fed only on honey and pollen. The queen-larvae, on the other hand, are always and solely fed on "royal jelly." They consume great quantities, their roomy royal-cell being flooded with it. This food has an extraordinary effect on the future of the brood; if continuously given to a larva of the worker-class that larva will develop into a queen-bee; if continuously given to a drone-larva, the resultant drone will be of an enormous and monstrous growth, but its testes will suffer a fatty degeneration and disappear. After five and a half days the queen-larvae, and after six days the drone and worker-larvae, cease to feed, and turn into pupae.

The termites or white ants, not to be confused with the true ants of our fields and woods, make enormous ant-heaps in the warmer parts of the world. They keep the interior of their home scrupulously clean. They eat all refuse, cast skins, dead bodies of their friends; even the matter which

FOOD OF INSECTS

has passed through the alimentary canal is eaten again and again until the last atom of nutriment has been abstracted and it has no further value as food. What is then excreted is either removed from the nest or used to plaster the pitch-dark corridors along which the insects are ever ceaselessly hurrying. Not only do they devour food which has come away from one end of the alimentary canal, but they frequently regurgitate food through the mouth to supply a hungry colleague. Should a termite fall ill or be in any way disabled, the weakling is devoured alive by other members of the ant-heap. White-ants, like certain tropical tree ants, are agriculturists. They cultivate certain fungi, planting the spores in appropriate places. They devour the mature fungi.

There is hardly anything that some insects will not eat and thrive on. The caterpillars of the *Tinea* moth feed on the hairs of fur which consist of keratin, a substance allied to protein, a very dry diet—on deers' horns, the hooves of horses, etc. Many bugs and plant-lice live upon sap; other bugs, flies, gnats, mosquitoes and lice suck the blood of vertebrates; and like most blood-sucking insects they take a large part in spreading diseases, amongst man at least. Mosquitoes infect human beings with malaria. Lice give rise to trench fever and typhus fever. Flies convey (most of them mechanically) the germs of typhoid fever. The blood-sucking insects are very persistent and very successful in attaining their object. I used to be told when I lived in Southern Italy that if you placed the legs of your bed in iron basins filled with water and placed the bed in the centre of the room you would escape being bitten by bugs, but this is not true. They crawl up the walls and along the ceiling and drop down on you. As a great American poet writes:

> The Lightning-bug has wings of gold,
> The June-bug wings of flame,
> The Bed-bug has no wings at all,
> But it *gets* there all the same!

The entrance to the mouth of the house-fly is so minute that it has to take in its food always in a state of solution. Solid particles cannot pass through it. Hence, when you see a fly feeding on a piece of sugar, you may be certain that it has excreted some saliva, and that this saliva has liquefied the sugar, and it is this sweet solution that the fly sucks into its body.

One of the most voracious of insects is the sacred beetle, *Scarabeus sacer*, which was regarded by the Egyptians as an emblem of immortality. This beetle, which exists all round the Mediterranean area, prepares a large ball of animal excreta much larger than itself. This it buries in a hole, and when it begins to eat it its appetite is stupendous. Fabre watched one of these beetles from eight in the morning till eight at night, and it never ceased eating, and its appetite was never satisfied until it had consumed its great sphere of food. At one end it was continuously eating and at the other continuously passing out undigested matter in the form of a long cord. Every 54 seconds this cord was added to, and in the course of the single meal the cord attained a length of three metres. In the course of a dozen hours this sacred beetle digested or at any rate passed through its alimentary canal an amount of nourishment very nearly equal to the volume of its own body.

Another voracious insect is the female mosquito, which will so gorge itself with blood that it cannot retain it. Whilst it is still pumping in blood at one end, blood is streaming out at the other end. As Dr Johnson said of the school where discipline was maintained "without recourse to corporal punishment," "But surely, Sir, what they gain at one end they lose at the other."

Many insects cause great damage to fabrics and food stores. The larva of the clothes-moth devours all sorts of woollen goods, carpets, furs, and sometimes silk. Stored flour and meal are devoured both by the Mediterranean flour-moth and the meal-moth, and their larvae occur in such quantities in flour-mills that the machinery is sometimes put out of action by them. Similarly, a considerable number of beetles of the genus *Tenebrio* consume all kinds of grain and cereals. The roof of the Great Hall of Westminster has been almost destroyed by the larvae of a beetle known as the death-watch, *Xestobium tessellatum*. There were holes in its gigantic beams so big that you could have put a child into them. The

FOOD OF INSECTS AND SPIDERS

Government have had to spend £10,000 a year for some years in making the roof safe. Many weevils destroy cereals or rice. The larvae of other species of beetle devour stored hams and bacon and lard. Others live on tobacco, and are so varied in their tastes that they will feed on raisins, cayenne pepper, dried fish, ginger, liqueurs and pyrethrum powder, which is the basis of all insect-powders. The adult tobacco beetle, which so often spoils cigars and pipe-tobacco, is about 1-16th inch in length, and it is curious to note that the size of the adult is always increased when the larva has been deposited in selected cigars of a very high quality. The female prefers the milder cigar to those that are stronger, a "Claro" rather than a "Colorado" and expensive Turkish tobacco rather than the Virginian. Altogether a beetle of a refined and delicate taste!

It should be noted here that in insects it is the larva who does most of the eating and all the growing. Most insects attain their largest size at the end of the larval period. They shrink a little as a rule when they become pupae or chrysalides and the adult insect weighs less than the last larval stage. Those insects that live but for a few hours, such as the Mayflies (EPHEMERIDAE), take no food during the adult stage. They simply pair and the male dies; the female lays her eggs and then she promptly dies too. During the intermediate stage, the pupa or chrysalis, no insect takes any food at all. Many insects devour green leaves and are one of the gravest dangers to crops and forest. Famine follows the locust, and of recent years the larva of a small moth, *Tortrix viridana*, has been destroying our oak-trees. Such leaf-eating insects derive their bodily pigments directly from the leaves they eat.

Spiders are in the main carnivorous. They take only liquid food, for like the house-fly their mouth is minute. They suck the fluids from captured insects or other spiders, etc. Although the mother often exercises a certain maternal care and shows an interest in her offspring, she is but a poor helpmate, as she frequently devours her husband as soon as they have paired. Scorpions again are carnivorous, living on spiders and insects which they seize with their nippers and sting to death with the product of a poison-gland situated at the end of the tail. Female scorpions also destroy their mates after pairing.

66 FOOD

They suck the blood and other juices of the male, for they also have minute mouths.

When we come to the vertebrates, we find their food is as diverse as is that of the Invertebrata. The ultimate food of fishes is the innumerable minute, unicellular, green algae, known as Diatoms. These organisms float near the surface of the waters, and in sunlight are enabled to build up their bodies from water and carbon dioxide dissolved in both sea- and fresh-water. Diatoms are eagerly devoured by incredible numbers of small CRUSTACEA, largely by the COPEPODA, and these minute Crustacea thus form a second stage in the food of fishes. Many of these Copepods are of surpassing beauty, as beautiful as birds-of-paradise or Mr Brock's fireworks. Yet they are hardly visible to the naked eye. The Crustacea are, as regards numbers, the predominant inhabitants of the sea, exceeding any other group of marine animals. Some of them live at the bottom of the sea and some on the sea-shore, and one small group of minute size, the OSTRACODA, act as scavengers, consuming the dead bodies of other animals. All form food for fishes. Below the limit of the

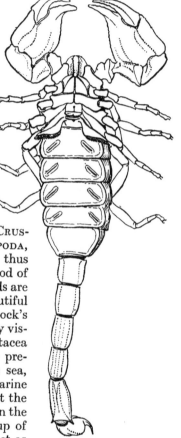

FIG. 19. Underview of an Indian scorpion, *Scorpio swammerdami*. From Shipley and MacBride. $\frac{2}{3}$ nat. size.

penetration of sunlight, which is at the very most about 300 fathoms—for the most delicate pan-chromatic plates register

FOOD OF FISHES

faint traces of attenuated light at this depth—green plants cannot live, and there the animals are necessarily dependent on each other for their food. They are all carnivorous, and the deep sea fishes have tremendous jaws with rows of vicious-looking teeth with which they catch their prey. Down in the depths of the ocean one realizes how right the Frenchman was who said that life was summed up in the conjugation of the verb "manger":

>Je mange,
>Tu manges,
>Il mange,
>Nous mangeons,
>Vous mangez,
>Ils mangent,

or its terrible correlative:

>Je suis mangé,
>Tu es mangé,
>Il est mangé,
>Nous sommes mangés,
>Vous êtes mangés,
>Ils sont mangés.

Sharks and dogfish are well known to be carnivorous, and the former often attack man. A good many fishes live upon their kin, though the cod varies its fish diet by eating crustacea, and the haddock chooses as its food such invertebrates as starfish, sea-urchins, and crustacea. The plaice and flounder and other flat fishes devour molluscs such as cockles, clams and mussels; and the sole consumes worms. The fishes which live chiefly on floating Copepods and small crustacea frequently have comb-like processes inside their gill arches. This is true of the basking shark, *Cetorhinus*, one of the largest of fishes, 40 ft. in length, and also of the herring. The processes act as a sieve, sifting out the crustacea from the water which enters the mouth and then flows out over the gills. A few fishes browse upon the seaweed growing round the coast, whilst sardines, the young of the pilchard, are also herbivorous, living upon the diatoms which they sift from the surrounding water much in the same way that the herring sifts its copepods, or the whale-bone whale its floating molluscs.

68 FOOD

Frogs, toads, and newts consume snails, slugs, insects, worms and almost any kind of small invertebrate animal. The prey of the frog is caught by the tongue, which is free behind but is fastened to the mouth in front. It is suddenly flicked out and drawn back with a fly or some other morsel of food attached to it. The tongue may be thrown out as far as an inch from the edge of the mouth and whipped back almost immediately with the prey curled in it. It is covered with a sticky secretion to which the prey adheres. Small animals pass straight into the gullet without touching the

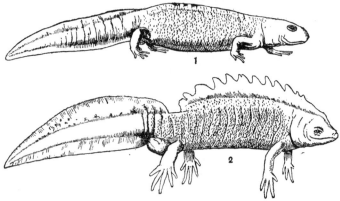

FIG. 20. The Warty Newts, *Malge cristata*. 1. Female. 2. Male at the breeding season with the frills well developed. (From Gadow.)

teeth, but large animals such as slugs and worms are caught between the teeth and the upper edge of the roof of the mouth. Frogs and toads have a curious way of making sure that their larger prey does not escape them. The orbit which lodges the eye has no bone at the bottom, so that the eye can descend into the cavity of the mouth. There is a muscle known as the "retractor bulbi," which pulls the eyeballs down, and when a worm is well within the mouth, this is done, and the eyes are used as clamps to prevent the worm from escaping, thus in a very literal sense they have "their eyes on their food." The young tadpole has no mouth at birth, only acquiring one after being hatched for a few days. It then begins

FOOD OF REPTILES

to feed upon the leaves of water plants and other vegetable matter. It is not till towards the end of the tadpole stage that it becomes weaned and adopts an animal diet. The change from the herbivorous diet of the tadpole to the carnivorous one of the young frog is accompanied by a remarkable shortening of the intestine. The length in the tadpole before metamorphosis is about 8–10 mms., that of the frog just metamorphosed is about 3–4 mms. As a rule an animal diet is more easily digested than a vegetable one, and the alimentary canal can be correspondingly shortened.

Of the four groups of reptilia, two are entirely carnivorous and two eat both plants and animals. Crocodiles and alligators are confined to the tropical or sub-tropical regions of our globe, and are almost exclusively fresh-water. They live entirely on flesh and in one order solely on fish; but most are, within their limits, catholic in their tastes, and eat any animal that comes in their way. They are not capable of swallowing large animals whole, and hence they have to mangle the body, tearing off bits by a series of sudden jerks.

Tortoises and turtles are found in the sea, in fresh water, and on land, chiefly in the warmer parts of the world. Some are herbivorous, such as the common tortoise one sees sold in the streets of our towns, and the gigantic land tortoises one sees in the Zoological Gardens. They feed on fallen fruit, leaves, or grass. Being of a placid disposition and almost devoid of emotions or passions they manage to get along on a meagre diet of a very unstimulating nature, and so live to an immense age. The mud-turtles are carnivorous, living on worms, insects, small fish, and even newts. The snapper-turtles, which are comparatively active and are capable of inflicting severe wounds on people who have invaded their waters, are more active than most of the group and enjoy a more heating diet. They will eat fishes, frogs, and water mammals, and seize and drown ducks and geese. The green-turtle so beloved of epicures feeds on seaweed growing among the sea-wrack which is found at the bottom of the ocean. Logger-headed-turtles live on cuttlefish and other molluscs, the shells of which their strong jaws crush with the greatest ease. The leathery-turtle, the largest of all its group, is entirely an animal feeder, killing and eating crustaceans, molluscs and fishes.

Snakes are entirely carnivorous, and their method of feeding is gruesome. They prefer to swallow their prey alive, having

FIG. 21. The Texas Rattlesnake, *Crotalus atrox*, reduced. From Stejneger.

put it out of action by crushing it or poisoning it. The snake then shifts its living victim about until it is in a convenient position to be swallowed head first, and it now proceeds gradually to drag its body over the prey. This process may take hours. Its jaws are so arranged that they can gape very

FOOD OF REPTILES AND BIRDS

widely, the two lower jaws being connected by an extensile elastic ligament. All the teeth are recurved, and first one half of the lower jaw is pushed forward and then the other, and gradually like a man climbing up a rope hand over hand the snake creeps over its food. When the snake has passed its mouth over the victim, the neck may be tremendously extended, every scale starting apart. In order to assist in swallowing there is a great discharge of saliva, and in order that the reptile should not be choked, the entrance to its air pipe is placed far forward and can be protruded between the two lower jaws while they are at work.

Some birds and mammals are immune to snake poisons and some snakes are harmless. The origin of the group is obscure, but the poison habit is certainly an acquired one and not inherited from their remote ancestors. Well, as Sancho Panza reminds us, "we are as God made us, and some of us even worse."

Lizards vary greatly in their diet, the majority feeding on insects, worms, and occasionally small birds and mammals; but some are herbivorous, and then the intestine is unusually long. Of the three English species, the young of the common lizard, *Lacerta vivipara*—the only reptile found in Ireland—feeds upon aphides and other minute insects, the adult eats larger insects and the sand lizard, *Lacerta agilis*, feeds on insects throughout its life. The slow-worm or blind lizard, *Anguis fragilis*, which is so often mistaken for a snake, is fond of slugs.

Birds, like lizards, are sometimes carnivorous, sometimes herbivorous, and sometimes both. They do a great amount of destruction in our gardens and orchards, and our cornfields, but the species of birds are so numerous and their habits are so diverse that it is difficult to generalize about them. As they have only a beak to help them to pick up their food, they prefer as a rule something easily picked up, such as seed, although of course the great vultures hold their victims in their claws and tear the flesh off with their beaks. Young grouse feed exclusively on insects till the end of their third week, when they become weaned and take to a diet of heather-tops with which by accident may be mingled a certain number

72 FOOD

of insects and small land molluscs. They thus reverse the policy of the frog. The thrush is fond of snails and it will break their shells by pounding them on "sacrificial" stones until they are all in pieces. The shell-fragments are carefully picked off, and the soft body of the snail disappears in one gulp. Certain sea-birds, such as petrels, have in their beaks a mechanism for straining off the smaller marine organisms

Fig. 22. The Duckbill, *Ornithorhynchus anatinus*.

from the sea-water which escapes at the sides of the mouth, just as in certain fishes.

Fledglings are notoriously voracious and their parents are kept pretty well occupied in catching caterpillars and other insect food for their young. In some cases this food is half digested by the parent and then regurgitated into the gaping mouths of their young.

Mammals are as varied in their food as birds, but they have one point in common. Their young invariably feed on milk,

MAMMARY GLANDS

the excretion of the mammary glands. The lowest order of mammals, the MONOTREMATA, lay eggs and the young hatch out in a very rudimentary condition. They are at once

FIG. 23. The Rock Wallaby with young in pouch, *Petrogale xanthopus*.
After Vogt and Specht.

transferred to a mammary pouch which has no teat, but from whose diffused milk-glands they can absorb milk. One of the members of this group, the Australian ant-eater, *Echidna*, feeds by thrusting its sticky tongue into ant-holes. The ants adhere to it and are soon swallowed. The second genus, the duck-billed platypus, *Ornithorhynchus*, is also carnivorous, living chiefly on grubs, worms, snails and occasionally mussels.

All other mammals have teats, and in the MARSUPIALIA, kangaroos, wallabies and wombats, etc., these are hidden away in the pouch or marsupium. This is a sac or cavity on the under or ventral surface of the body, into which, as the uninformed school-boy said, "the kangaroo retreats in moments of extreme danger." The young are born in a very rudimentary state and are quickly transferred by the lips of the mother to the pouch, and permanently attached to the teats. The young of even the larger kangaroos are not much larger than the little finger of a man. Kangaroos and wallabies are herbivorous, browsing on grass and herbage and occasionally roots. Others, like some of the phalangers, are omnivorous. As its name indicates, the Tasmanian wolf is carnivorous, and causes considerable harm to the stock-keepers' herds. Opossums are largely insectivorous, but will also eat carrion, eggs and lobsters.

The kangaroo is regarded as the typical and national animal of Australia:

> Kangaroo! kangaroo!
> Thou spirit of Australia,
> That redeems from utter failure,
> From perfect desolation,
> And warrants the creation
> Of this fifth part of the earth;
> Which would seem an after-birth.
>
> BARRON FIELD.

A curiously diverse group is that of the EDENTATA. It contains the sloth, ant-bears, and armadilloes. Like the Australian ant-eater, the ant-bear consumes nothing but white-ants (*termites*) and true ants with their larvae. It uses its great claws, with which it can successfully defend itself from many enemies, to tear down the ant-heaps and hills,

FOOD OF EDENTATES

often 10–20 feet in height and as tough as baked clay. The armadilloes, which are so well protected by their plated armour, live in pairs and generally appear only at night. They

Fig. 24 Tamandua Ant-eater, *Tamandua tetradactyla*.
From Proc. Zool. Soc. 1871.

devour carrion, but they also eat insects and vegetable fibres. Those curious animals the sloths, from South and Central America, live upside down, hanging on with their strong,

Fig. 25. The Six-banded Armadillo, *Dasypus sexcinctus*.
After Vogt and Specht.

curved claws to the branches of trees. "They live suspended and sleep suspended, and in fact pass their whole life in a state of suspense, like a young curate, when he is distantly related to a bishop," as Sydney Smith tells us. They live upon leaves, young shoots and fruit, the juices of which supply them with sufficient fluid, for they never drink. The Tamandua ant-eater, from the same regions, is also arboreal.

The mammals that live in the sea are the manatee and dugong (SIRENIA), the whales (CETACEA) and seals (CARNIVORA). The last two are carnivorous. Whales fall into two great groups, whalebone whales and toothed whales, the latter including the dolphins and porpoises. The first group have enormous mouths, from the top of which hang down plates of whalebone fringed on the inner edge. These whalebone plates may be as long as twelve or even thirteen feet. As they swim through the water, innumerable floating organisms, including large numbers of surface molluscs known as PTEROPODA, become entangled in the fringes, and these form the food of this great marine monster. The spermaceti whale has a hollow at the top of its skull, and its head is swollen out by a huge mass of a fatty nature known as spermaceti. This produces oil of a very rare quality which is necessary for lubricating the parts of extremely fine machinery. No other oil can replace it, and when whales become extinct, as they threaten to do in the course of time, unless some substitute can be found, fine machinery will be in a poor state. The gigantic head of this whale so overweights its jaws that this species is said, like the shark, to turn over on its back when it wishes to attack its prey. Some toothed whales live largely on cuttle-fish, which they find at the bottom of the sea. Many of these gigantic molluscs are only known from portions that have been vomited up by wounded whales, whose skin is often scored half an inch deep by the powerful underhung parrot-like beak of the cuttlefish. This sperm-whale has frequently in its stomach concretions which are known as ambergris. They seem to be formed around the beaks of cuttlefish. Ambergris is a substance of extreme value, for though it has but little perfume of its own, it is used in fixing and improving perfumes with which it is mixed.

Dolphins are sometimes extremely ferocious, attacking seals, porpoises and even larger whales. The bodies of thirteen porpoises and fourteen seals have been taken from the stomach of a great killer-whale or grampus. These "wolves of the sea" congregate both in spring and autumn around the Pribylov Islands in the North Pacific, to feed on the fur seals which breed there. The destruction caused by them is very great. The stomach of one killer contained eighteen seals and that of another twenty-four; each had consumed fur of the value of some £400 to £500 according to the sealers' estimate. Both killer-whales and seals annually migrate southward, but what happens then is unknown.

> You mustn't swim till you're six weeks old,
>   Or your head will be sunk by your heels;
> And summer gales and Killer Whales
>   Are bad for baby seals. KIPLING.

There is a special genus of dolphin which inhabits the Cameroon River, and it alone amongst the Cetacea is said to be herbivorous, living on water weeds.

The SIRENIA or Sea-cows (the Dugong and Manatee) are as aquatic as the whales and apparently never come on land. They live on sea-weeds and water-plants. They pass their life in the shallow waters of the tropics and are common about the mouths of rivers, which they often ascend. For the most part they live a placid existence under water with their body arched, their tail and head being bent in beneath the body. Of three specimens observed in an aquarium, a male and a female and a young one, the male came to the surface more frequently to breathe than the female. The young specimen remained below the surface on an average four and a half minutes, with a maximum of nine and a half minutes, whereas the larger specimens would remain submerged for periods of twelve minutes and on occasions for sixteen minutes. Thus the Sirenia like the Cetacea are very greatly specialized for an aquatic life.

The great order of UNGULATA includes the two small groups of the HYRACOIDEA or conies and the PROBOSCIDEA or elephants and two large groups, the PERISSODACTYLA or the odd-toed ungulates, horses, tapirs and rhinoceroses, and the ARTIODACTYLA, the even-toed ungulates, hippopotamuses, pigs, camels, deer, giraffes, sheep and oxen. These are the greatest source of food for the terrestrial carnivorous animals, which we shall consider later. In the main it is in the ungulates that the vegetation of the earth is converted into forms of food which carnivorous mammals can digest and assimilate. The Artiodactyla stand somewhat apart from the other three groups and probably have a different origin in the remote geological past.

The little *Hyrax*, the coney of the Bible, is externally much more like a rodent than like an ungulate, but anatomically it is allied to the latter group. Conies are confined to Arabia, Palestine and Africa. They feed by night, their principal diet

FIG. 26. Indian Rhinoceros, *Rhinoceros unicornis*. From Wolf.

being the leaves and young shoots of plants. Their nearest allies the elephants, again, are strictly herbivorous; any visitor to the Zoo will recall their fondness for fruit, cakes and buns. Their extremely delicate and sensitive proboscis is well adapted for seizing and conveying food to the mouth.

When we come to the odd-toed ungulates, the PERISSODACTYLA, we find that amongst them are many highly domesticated animals such as the horse and the ass, whose food in the main is fresh or dried grass. But the wild-ass of Central Asia will consume the woody plants which form the main vegetation of that very arid region. As a rule the rhinoceros rests by day. Towards evening it will turn out to feed, living

FOOD OF UNGULATES

chiefly on young shoots and young branches of acacia, varied with fruits and roots. Rhinoceroses cause great damage in sugar-cane and melon fields, and are especially dreaded by the owners of cacao plantations, where they do much harm to the young growth. The tapir, like the rhinoceros, shelters itself during the day time and feeds at night on palm leaves or fallen fruit or swamp-grass and water-plants.

The ARTIODACTYLA or even-toed ungulates are equally herbivorous, but the pig is omnivorous and will eat flesh, fresh or decayed, and any amount of filth. Wild pigs are often very destructive to crops, but away from human activity they will often feed upon roots, sedges, and carrion, and excreta. They trample and root up the soil in a terrible fashion and destroy all vegetation. The hippopotamus, owing to its gigantic bulk, consumes an immense amount of food. Its stomach is enormous, measuring 11 feet in length and capable of containing from five to six bushels of foodstuffs. It also causes considerable damage to rice-, millet-, and sugar-plantations, causing even more damage by its trampling than by its feeding. Away from human habitations "hippos" live chiefly on water plants. The llamas of South America feed by day and have to descend from the rocky heights of the Cordilleras, to which they resort as summer approaches, to obtain food in the valleys. In their natural state their allies, the camels of Africa and Asia, feed on green food, though, when domesticated, grain is largely given them; but to keep them in health green food is essential. It is often supplemented by dates. Like the ass, they are fond of eating and swallowing the most thorny plants.

Deer browse on grass, but they also eat the leaves of trees and shrubs. The wapiti of North America is rather a coarse feeder, and will readily take food rejected by horses or oxen. Reindeer are apt to migrate from the inland to the coast during the autumn and to vary their ordinary diet of moss by eating sea-weeds. They have a specially modified horn with a flange for shifting the snow so that they can get at the underlying moss. The musk-ox of Arctic Canada is more closely allied to the sheep and goats than to the ox. It lives in herds. The caribou or American reindeer devours leaves, grasses, and aquatic plants, but its chief source of food is lichens. The long neck of the giraffe enables it to crop the leaves of the local trees, which are picked one by one with the help of its long flexible tongue. Goats will eat almost anything, but when confined to their natural rocky habitat they live on the scanty mountain herbage, which consists largely of lichens. The food of sheep and oxen is too well known to dwell upon, but the amount consumed in the course of a year by a milking cow must be enormous. A

FIG. 27. The Musk-Ox, *Ovibos moschatus*.

first-class milker will produce in a year an amount of milk which weighs fifteen times her own body weight. That cows also eat the leaves of trees is shown by the fact that the distance between the ground and the horizontal flat level of the leafy part of trees is equal to the height a cow can raise her mouth. The stomach, which is divided into four compartments, is so arranged that the food which has found its way into the first division (rumen) is sent up the oesophagus into the mouth again, where it is solemnly chewed over and over again. It is an interesting point to notice that during this act, which is called *rumination*, the lower jaw moves side-

FOOD OF RODENTS

ways once either to the left or to the right, and the whole of the rest of the time it moves in the other direction. The ungulates are the source of food to the carnivorous land animals, and form a large part of the flesh consumed by men. "The charred limb of a ruminant" is a favoured dish of mankind.

The RODENTIA, hares, rabbits, rats, mice, beavers, squirrels, etc., gnaw, and they will gnaw pretty well anything. They are very easy to find amongst mammals by the presence of a pair of large chisel-shaped front teeth which grow con-

FIG. 28. The Musquash, *Fiber zibethicus*.

tinuously. They are almost entirely herbivorous and they get their food by gnawing. They are in the main terrestrial and they burrow; but a few are fresh-water and a few—like the flying squirrels—are arboreal. The musquash of North America burrows in banks and feeds, chiefly at nights, on water-plants.

The carnivorous animals, CARNIVORA, as their name implies, feed both on flesh-eating animals and plant-eating animals, but ultimately they are dependent on the herbivora. Some of them, such as the hyena and the jackal, feed on carrion. The mongooses attack and devour live snakes, even the most venomous. They have been introduced into certain West Indian Islands with a view to keeping down snakes, but the introduction has not been entirely a success. The mongoose seems to escape the fangs of the snake by his unusual activity, and possibly he is slightly immune from snake poison. One species, however, lives almost entirely on frogs and crabs.

FIG. 29. The Common Skunk, *Mephitis mephitica*.

The skunk with its vile smell is valued for its fur and there is now more than one skunk-farm in England.

The following account of the food of a fox is taken from Miss Frances Pitt's *Woodland Creatures*:

Wherever we meet with the fox it is a wild animal, fearing and shunning man and all his works, a hunter of rabbits, birds, and mice, a raider of poultry-yards, and sometimes in mountainous districts a slayer of young lambs. It is rarely that the lowland foxes commit the latter crime, it takes an old hill fox to do it. The average fox of our

FOOD OF CARNIVORA AND INSECTIVORA

English woods lives on much smaller fare, rabbits being the principal item in its menu, as can be proved by examining the droppings, which are invariably full of rabbit fur. From the same source will be obtained evidence that many other things are not to be despised, down even to grubs and insects. It is often astonishing the number of beetle wings, or rather wing-cases, that will be found in the excrement, the hard elytra having passed through undigested. The fact is that a fox will eat many unexpected things, from beetles, frogs and fish, to even fruit. It has a liking for sweet things, and I knew a tame fox that would do anything for jam. There is undoubtedly some foundation for the fable of the fox and the grapes. Foxes are also very fond of mice, in particular the short-tailed meadow voles, which are so plentiful in long grass. They will watch for and pounce upon them, often killing numbers; indeed, the successful stalking of field mice seems to be the first step in the education of the cubs, when they begin to learn their profession as hunters.

Of the marine carnivora the seal eats fish, but its ally the walrus, of which there is only one species, and that only found around the Arctic regions, consumes molluscs, as readers of *Alice Through the Looking-Glass* will recall:

> "O Oysters, come and walk with us!"
>   The Walrus did beseech.
> "A pleasant walk, a pleasant talk,
>   Along the briny beach:
> We cannot do with more than four,
>   To give a hand to each."
>
> "I weep for you," the Walrus said:
>   "I deeply sympathize."
> With sobs and tears he sorted out
>   Those of the largest size,
> Holding his pocket-handkerchief
>   Before his streaming eyes.

As their name again implies, the INSECTIVORA, hedgehogs, shrews and moles, live on insects, but the mole and the hedgehog will eat leaves, and, as is stated above, the former will consume its own weight of earthworms in the course of twenty-four hours. The desman lives in burrows, in banks and feeds on water insects. Its hind feet are webbed and its tail is flattened. Water-shrews, common enough in England and Scotland, will feed on the flesh of larger animals when they find them dead.

The CHEIROPTERA, bats, resemble Insectivora which have taken to the air. Many live on insects, but some of the vampire bats suck blood, and as blood requires little digestion their stomach is extremely simplified. The members of one of the two sub-orders subsist entirely on fruit. One species

FIG. 30. Russian Desman, *Myogale moschata*.

of *Xantharpyia* lives in pyramids and tombs and is reproduced often in old Egyptian frescoes.

What we are accustomed to consider the highest order of mammals, because it includes man, consists of animals which are usually omnivorous. Chimpanzees, which are as a rule fruit-feeding animals, will in captivity take readily to meat. The gorilla feeds chiefly during the daytime on wild fruit, but will readily drink milk. The long-armed gibbons take readily to small birds, insects, spiders, and eggs, varying their

FOOD OF APES

Fig. 31. Female and young of *Xantharpyia collaris*. From Sclater.

diet with fruit and leaves. The Orang-utan from Sumatra and Borneo lives on leaves and shoots but chiefly on fruits. It builds a platform or nest on which it sits. Monkeys have a very varied diet, but in their wild state they live chiefly upon

86 FOOD

berries, seeds, fruit, and buds of trees. Lemurs are said to be fond of honey, but their chief diet is fruit and insects.

Both the teeth and the alimentary canal of man point to a mixed animal and vegetable diet, though this is modified by climate. In the cold of the Arctic the Esquimaux revel in

Fig. 32. The Orang-utan, *Simia satyrus*, sitting on its nest.
From a specimen in the Cambridge Museum.

blubber, whilst there are large tracts of country in the tropics where the natives live on rice, maize or bananas. Man is the only animal that cooks its food, but the distance between the rough cookery of the Australian native or an Abyssinian and that of a French chef is almost immeasurable. The higher branches of cookery are one of the chief arts in which men invariably rise superior to women.

CHAPTER IX

DIGESTION

ALIMENTARY CANAL—BODY-CAVITY—DIGESTION—
HORMONES AND VITAMINS—CALORIES—APPETITE

"The process of digestion, as I have been informed by anatomical[1] friends, is one of the most wonderful works of nature. I do not know how it may be with others, but it is a great satisfaction to me to know, when regaling on my humble fare, that I am putting in motion the most beautiful machinery with which we have any acquaintance. I really feel at such times as if I was doing a public service. When I have wound myself up, if I may employ such a term," said Mr Pecksniff with exquisite tenderness, "and know that I am Going, I feel that in the lesson afforded by the works within me, I am a Benefactor to my Kind."

Mr Pecksniff. *Martin Chuzzlewit*. CHARLES DICKENS.

ALIMENTARY CANAL

THE simplest of multicellular animals, METAZOA, such as the sponges, are compact but honey-combed by channels through which water with food suspended in it circulates. They have no mouth or stomach. They are all aquatic and their food is taken in by the cells exposed to the water. These cells behave like so many *Amoebae*. A little higher up in the scale, in certain TURBELLARIA, there is a mouth and a stomach, like the inner, closed tube of a thermos-flask. Each of the cells lining this cavity eat up the food-particles in an amoeboid manner. In the parasitic flukes, TREMATODA, the stomach branches so that the food is brought into close contact with all the cells of all the tissues of the solid body. In such animals undigested food, if any, passes out through the mouth. The larvae of bees feed on a perfect diet, "pap" or "royal jelly" at first, and then after four days a little honey is added,

[1] In Mr Pecksniff's time the physiologist had hardly emerged from the anatomist, and what little physiology was taught in our Hospitals and Universities was taught by the Professors of Anatomy.

88 DIGESTION

but all is so digestible that no undigested food is found in them, there is nothing to soil their waxen cells, and economy and sanitation march hand in hand.

In animals such as we are now considering the alimentary canal is embedded in the otherwise solid body and moves as the body moves. If the latter stretches, the digestive system stretches; if it contracts, the alimentary canal also contracts. The latter may branch throughout the solid tissues of the animal, but the longest branch is never longer than the animal.

BODY-CAVITY

In higher animals, however, a second cavity arises, the *body-cavity*. This is bounded on the outside by the skin and underlying muscles and on the inside by the muscular walls of the alimentary canal. The body-cavity is traversed by the stomach and intestine, and it houses, besides, the large glands connected therewith, the liver and pancreas, also the heart, the lungs, and the kidneys. Although in many of the simpler animals, *e.g.* the earthworm, the alimentary canal passes straight from

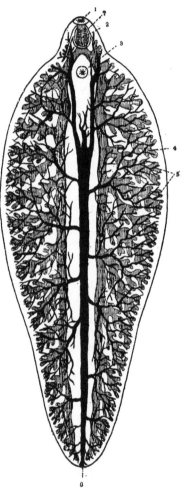

FIG. 33. Diagram of digestive and excretory system of the Liver-fluke, *Distomum hepaticum*. Magnified about 8. From Leuckart. 1. Mouth. 2. Pharynx. 3. Reproductive pore. 4. Branch of alimentary canal. 5. Branches of excretory system. 6. External opening of excretory system. 7. Nerve-ring.

THE BODY-CAVITY

mouth to anus and is the same length as the body, in more complex forms the existence of this body-cavity enables the intestine to grow to a length greatly exceeding that of the body. The intestine of man is just under 31 feet in length, five or six times the length of the body. Thus the digestive and absorptive capacity of the alimentary canal is greatly increased.

But the existence of a body-cavity has a further advantage. It frees the alimentary canal from the control of the surrounding tissues. It is no longer dragged hither and thither as the contractions of the surrounding tissues sway it. The intestine in a body-cavity can exercise movements of its own, movements independent of the surrounding body-wall; and this freedom of action is rendered possible by the existence of a space—the body-cavity—in which the intestine lies free.

Animals with an alimentary canal open at both ends—and they are the vast majority—might also be compared with a thermos-flask, if only the inner tube of the flask were fused to the outer tube at the lower end as it is at the top end and opened externally. Between the two tubes of the flask is a space, a vacuum in the thermos-flask, which represents the body-cavity of an animal. This space in animals is almost completely filled up with various organs, muscles, excretory and reproductive organs, glands, etc. If you were to put a foreign body such as a pebble into the inside or *lumen* of the inner tube of the thermos-flask, it might still in a sense be regarded as outside the body, the body being in reality the tissues which take the place of the vacuum of our flask. Now the food that passes from the mouth down the gullet into the stomach and thence into the intestine is also in a sense outside the body, and it has to be changed in many ways before it can soak through the lining of the intestine in the higher animals into the blood and be carried by the blood to the various active and hungry cells that constitute our body.

Animals eat every kind of food, and in all of those foods which are nutritious we find certain foodstuffs. In building up their proteins animals must have nitrogen compounds at least as complex as amino-acids. Nitrates are excreted unchanged. You might fill up the bodies of the unhappy, starving Russians with nitrates but not for one moment would the famine degeneration be arrested. Carnivorous animals eat meat, which contains proteins and fats. The proteins and the fats cannot pass through the walls of the intestine until they have undergone a change. Animals that are herbivorous consume a great quantity of starch and sugar, and with the starch is always more or less protein matter. Starch is again insoluble in water.

Lining the alimentary canal is a membrane, and it is only certain substances that can diffuse through this membrane. Sugar is soluble in water, and can pass through it, leaving the cavity of the intestine and arriving in the blood-vessels which run in its wall. Besides the foods mentioned above we take up a certain amount of salts and a certain amount of other mineral substances, most of which are capable of passing in solution through the membrane. We live then on protein matter, fat, starch or sugar, with a certain amount of other minerals and water. We must have proteins, for they alone contain nitrogen, and nitrogen is wanted to build up the new proteins formed in the body. We may fill our stomach with starches, proteins and fats, but we should starve unless these foodstuffs were rendered soluble. Until this is done the food is of no more use to the body in the alimentary canal than it was before it was swallowed.

But with the meat and the fat and the starch and the sugar there are a number of other substances which form, as it were, the packing of these foodstuffs. They are useless as food, and when once the food is dissolved out from them they pass away from the body as utterly indigestible. They pass away without being dissolved into the blood and have never formed any real part of the body at all. It seems that for the proper action of the intestine a certain amount of indigestible food such as cellulose, *i.e.* the coating of vegetable cells, and the tougher fibres of meat, is necessary in order to supply a certain amount of bulk without which the normal action of the intestine is liable to stagnate.

Digestion

Digestion is the act of preparing and dissolving the foodstuffs so that they may pass through the membranous wall

DIGESTION

of the alimentary canal and get into the blood, and by the blood be distributed to all the cells of the body.

There are a number of glands opening into the alimentary canal; the salivary glands, which produce saliva, open into the mouth; then there are innumerable glands packed close together in the walls of the stomach, which secrete the gastric juices. Then we have the liver and pancreas. The saliva passing into the stomach turns starch into sugar before the action of the gastric juice has rendered the stomach-contents acid. But saliva also helps to moisten the food and thus to make it more easy to swallow. When the Red Queen in *Alice Through the Looking-Glass* after their tremendous race offered the dry-mouthed and tired-out little girl a dry biscuit, one can well imagine that Alice had difficulty in swallowing it.

The gastric juice in the stomach dissolves up proteins and makes of them a solution which can pass through the membrane of the intestine. The bile which comes away from the liver, and also the pancreatic juice, break up the fat into the minute particles such as can be seen if milk be examined under the microscope. In technical terms they *emulsify* it. It is then capable of passing through the membrane which forms the coat of the intestine. But the emulsified fat, now termed *chyle*, does not pass into the blood directly. In the first place it passes into some small vessels called *lacteals*, and these lacteals are always filled with the milk-like solution of fats. Ultimately they combine into larger and larger vessels, and the largest of these pours the chyle into one of the great veins near the neck, so that in time the fat does get into the blood, and thence to all the tissues and cells that want it. Pancreatic fluid also helps in changing starch into sugar. Water passes into the blood along the whole course of the alimentary canal, and with it the dissolved minerals, so that all real foodstuffs ultimately pass either straight into the blood or through the lacteals into the blood, leaving behind in the intestine the undigested insoluble parts to be cast out of the body.

The blood—in higher animals—transports the digested food throughout all the tissues and in such a form as to supply the working active cells with nutriment which enables them to carry on their several functions, and—in growing tissues—to increase in size and to multiply the number of their cells.

Hormones and Vitamins

When the food taken by the mouth reaches the stomach and intestine, the digestive juices are poured out to meet it. The action is started by means of hormones, which in some way stimulate the secretory cells. The name hormone is derived from ὁρμάω, "I arouse to activity." These substances act as chemical messengers: when the food mixed with gastric juice enters the small intestine, a substance called "secretin" is set free and passes by the blood stream to the pancreas. Its action on the pancreatic cells is to cause the pouring out of the pancreatic juice into the intestine. It has been proved that the agent which causes the liberation of "secretin" is the acid of the gastric juice. An acid extract of intestine, when injected into an animal, will cause a copious flow of pancreatic juice even though the animal has taken no food. Hormones are produced by numerous glands in the body and exercise great influence on many processes, such as metabolism, respiration, growth, and the sexual characters. In only two instances, namely those of adrenaline and thyroxin, have hormones been isolated in a pure state and their chemical composition determined. They have both a comparatively simple structure, and there is reason to believe that other hormones are not of great complexity. Many are not destroyed by boiling, but the fact that they are present in such minute quantity and are adsorbed on to other substances renders their isolation exceedingly difficult.

The hormones of the body are arranged under a complex system. There appears to be co-operation and antagonism between different groups. The over production of one or the deficiency of another will upset the balance and cause far-reaching changes. Many disorders are due to a disturbed hormone balance. What knowledge we have of them has been gained by the removal of some of the glands from animals, or injection of extracts of glands. The symptoms produced can

VITAMINS

then be compared with those occurring in certain diseases of man. A frequent practice in medicine is the administration of gland extracts in the case of deficiency, and removal of parts of glands which have been secreting to excess.

But even if the food an animal eats is of the right quantity and quality it will not be assimilated unless certain obscure ingredients known as *Vitamins* are present. Nobody has succeeded in isolating a vitamin, though as knowledge has increased we have been able to get them in more and more concentrated forms. You can go into the chemist's shop and ask, as the lady did, for "half a pound of pure vitamins," but the chemist will be unable to supply you. In some way these unknown bodies exercise some control over nutrition. Without one form of vitamins children will not grow. The absence of another leads to scurvy in adults and children. Scurvy, which plays so great a part in pirate stories and in the narratives of Polar explorers, has for a long time been treated by administering fresh vegetables and lemon-juice or fruit, and it is these fresh foods which supply the vitamins. Some authorities associate the disease of rickets so common in childhood with an absence of vitamins, and much work on this subject has been and is being done since the war amongst the starving children of Vienna. Rickets can be cured by supplying the child with vitamins or by exposing it to the sun-light or ultra-violet rays; whether the rays cure of themselves or whether they promote the welfare of the vitamins is not yet clear.

Vitamins are classified as:
      Vitamin A (Fat Soluble).
      Vitamin B (Water Soluble).
      Vitamin C (Water Soluble or Anti-Scorbutic).

It has recently been shown that Vitamin A—which is possibly identical with the yellow pigment carotin—is "activated" by sunlight.

The chemical composition of vitamins is not known. They exist in extremely small quantities in various foods and cannot, as far as is at present known, be synthesized by animals. The animal is dependent upon the plant for them. It seems that their formation may be dependent on the photosynthetic process in the green leaf, the process which results in the formation of chlorophyll. Vitamin A at any rate only appears when the plant turns green. Vitamin C, however, appears when the seed germinates and before any green parts of the plant are formed, and Vitamin B exists in large proportions in the colourless yeast cells. Possibly the vitamins act as Enzymes (or ferments), pulling the trigger, as it were, to set other food substances at work to build up flesh and renew waste in the animal body. Possibly the same is true also in the bodies of some plants, *e.g.* Fungi and some Algae. Fat Soluble Vitamin A is found in many animal fats, *e.g.* in butter, but not in lard; and not in most vegetable oils or fats, though there is some in nuts and nut-butter. Lack of this vitamin may cause rickets in children. The bone does not harden, and growth is either not possible, or is badly stunted. Eggs and milk contain Fat Soluble Vitamin A, and in experiments on rats it has been found that rats fed on foods without this vitamin became almost paralyzed and could not move their hind legs, but revived when a little fresh milk or egg was added to their diet. Their eyes, which had become affected and their lids reddened, also became normal again. This Vitamin A is destroyed by oxygen, and this may explain its deficiency in lard, which is often exposed in shallow pans to the air. Margarine, or vegetable fat, does not contain Fat Soluble Vitamin A. Hence the Government Order that margarine (at least during the war) must contain a percentage of animal fat.

In investigating the causes of scurvy or scorbutic disease at the Lister Institute, it was found that growth in young mammals is also dependent on the presence of the Water Soluble Vitamin B in the diet given, and without the third Water Soluble Vitamin C the skin is not sufficiently nourished and scurvy ensues.

Water Soluble Vitamin B is found largely in the germ and aleurone-layer below the skin or husk in cereals or grain foods, *e.g.* in wheat, rice, maize. Two "deficiency" diseases, *Beri-Beri* and *Pellagra*, have been traced to the lack of this vitamin. Beri-Beri is largely a tropical disease common, amongst other places, in India, since machine-milling for rice was introduced. This process took away the germ and the

VITAMINS AND CALORIES

aleurone-layer and polished the rice, but the mechanical process so diminished the food-value of the rice as to cause disease. The preservation or retaining of the germ and aleurone-layer at any rate to some extent remedies this defect.

Pellagra is found where maize is a main article of diet, *e.g.* as polenta-flour in Italy and parts of America. It is caused also by machine-milling and the lack of the germinal part of the cereal. In the older hand-milling this was not lost. Turnip juice may supply the deficiency thus brought about.

The Anti-Scorbutic Vitamin, Water Soluble C, is found chiefly in green vegetable food, *i.e.* leaves of the cabbage, lettuce, clover, etc. It is also found in lemon-juice, but not in lime-juice. Our troops in Mesopotamia suffered badly from scurvy although supplied with quarts of lime-juice, but when given lemons instead the disease became much less prevalent. The old faith in scurvy-grass or *Cochlearia* as a remedy is justified by modern discovery.

Orange-juice is a valuable source of Water Soluble Vitamin C, and babies fed on a milk diet may safely be given four to five teaspoonfuls of orange-juice daily. A nursing mother should also take it, or some food supplying it, *e.g.* grape-juice or lemon, as the child may suffer from its deficiency in the mother's diet.

Growth is dependent on the presence of both Fat Soluble A and Water Soluble B in the diet, and they are necessary, though not perhaps in so high a percentage, in the diet of adults to sustain life and increase resistance to disease.

There seems to be something—not yet isolated—in sea-water which acts like the vitamins. The average composition of the salt water is accurately known and is very fairly constant. Sea-water can be artificially prepared, but in this purely artificial medium marine organisms will not grow. After adding a trace of natural sea-water (1 to 4 per cent.) to the artificial fluid all sorts of marine organisms flourish abundantly in the mixture. Something akin to vitamins must have been introduced.

CALORIES

Since the processes that go on in living animals and plants involve slow combustion, it is necessary to form some estimate of the quantity of heat provided by the food, it being assumed by the principle of the conservation of energy that, although it can be transformed, the original quantity of energy remains constant. Physicists have chosen, to enable them to measure a quantity of energy of any kind, a certain amount of heat as a unit. This is known as a calorie, and a great Calorie is that quantity of heat which is necessary to warm one kilogramme of water from zero to one degree Centigrade. Heat was chosen as the unit of measurement, for heat alone holds a peculiar position in relation to other forms of energy. It is the sole form into which all others can be completely transformed. But it is now doubted whether complete transformation can take place even into heat.

People who are interested in diet have found out how many units of heat, called Calories, must be supplied by the food in order to keep the animal in health. A three year old child requires 1000 large Calories a day, a fourteen year old child about 1800, a full grown man about 3000 to 3500 according to the work he is doing, whilst a woman does quite comfortably on 2100 calories. But you can have a theoretically ideal food with the requisite number of Calories and yet, if it be unpalatable or lacking in vitamins, you cannot induce people to eat it or their bodies to assimilate it. Well-to-do people in civilized nations generally eat too much, and they generally have a mixed diet, whereas some of the Red Indians of North America are almost wholly flesh-eaters; the enormous populations of China, India and Japan are in the main vegetarian; certain tribes in Central Africa live entirely on bananas; Esquimaux live largely on blubber and on fish. But there is little doubt that the more civilized and more progressive races are omnivorous. The teeth and digestive tract of man are characteristic of an omnivorous animal, and this is equally true of bears.

Appetite

"You may take a horse to the water, but you cannot make him drink." You may have a most scientifically prepared food, containing the proper amount of Calories and an adequate supply of vitamins, but, unless it is palatable, people

APPETITE AND HUNGER

will not eat it. Appetite is hunger writ small. Hunger undoubtedly induces people to eat food which a well satisfied man would not care to tackle. On the other hand, elaborate sauces and good cooking will often tempt a jaded appetite to put forth new efforts. The savour of cooked food is always provocative of appetite. It induces the saliva to flood the mouth and the gastric juices to flow into the stomach. As a great Russian physiologist has said, "Appetite is juice."

"One man's food is another man's poison," and although the lower animals keep to a fairly steady and monotonous diet, man will ransack the world for delicacies. The part that the spice trade played in international politics during the Stuart and later times is an example of the power that appetite has over the affairs of man.

Sydney Smith used to say that a widely spread view of Heaven was "eating *pâté de foie gras* to the sound of trumpets." This no doubt inspired Lord Beaconsfield, whose epigrams usually had an ancestry, to utter the wish "O, may I die eating Ortolans to the sound of soft music!" Charles Lamb favoured a simpler diet. He held "that a man cannot have a pure mind who refuses apple dumplings," and although he was not particularly fond of vegetables, he observes that "asparagus seems to inspire gentle thoughts." But where he really let himself go was over roast sucking-pig:

See him in the dish, his second cradle, how meek he lieth!—wouldst thou have had this innocent grow up to the grossness and indocility which too often accompany maturer swinehood? Ten to one he would have proved a glutton, a sloven, an obstinate, disagreeable animal—wallowing in all manner of filthy conversation—from these sins he is happily snatched away—

Ere sin could blight or sorrow fade,
Death came with timely care—

his memory is odoriferous—he hath a fair sepulchre in the grateful stomach of the judicious epicure—and for such a tomb might be content to die.

Dr Johnson was devoted to the pleasures of the table. Boswell records that he said:

Some people have a foolish way of not minding or pretending not to mind, what they eat. For my part I mind my belly very studiously, and very carefully; for I look upon it, that he who does not mind his belly will hardly mind anything else.

He was a voracious feeder and rather a coarse one.

> When at table, he was totally absorbed in the business of the moment; his looks seemed riveted to his plate; nor would he, unless when in very high company, say one word, or even pay the least attention to what was said by others, till he had satisfied his appetite, which was so fierce, and indulged with such intenseness, that while in the act of eating, the veins of his forehead swelled, and generally a strong perspiration was visible.

Mrs Piozzi records that "his favourite dainties were—a leg of pork boiled till it dropped from the bone, a veal pie with plums and sugar, and the outside cut of a salt buttock of beef," and it is recorded of him that on one occasion he emptied the butter-boat full of lobster sauce over his plum-pudding.

But I must stop dealing with these literary appetites, and get on with the next chapter.

CHAPTER X

RESPIRATION

PLANT RESPIRATION—RESPIRATORY ORGANS IN ANIMALS—
DUST AND SAND—HAEMOGLOBIN—ANAEROBES

> The breath of Heaven fresh blowing, pure and sweet,
> With dayspring born; here leave me to respire.
> MILTON, *Samson Agonistes*.

PLANT RESPIRATION

WITH the inevitable exceptions, which will be considered later, all animals and plants breathe, that is to say, they take in oxygen and give out carbon dioxide; and unlike the taking in of food, which is intermittent, respiration goes on all the time. It is true that in green plants in daylight the process is veiled by the fact that they take in as part of their food a good deal more carbon dioxide than they breathe out, and as a result of the metabolism or the breaking up of the complex substances which build up the cells and the tissues they excrete more oxygen than they take in through breathing. We have in plants two systems working "contrariwise," as Tweedledee would say: (1) the breathing, which works day and night in the light or in darkness; (2) the feeding, which works only in green plants in daylight.

Life is a slow combustion: just as a steam engine is enabled to move about and maintain a high temperature by the combustion of coal, which requires the presence of oxygen—for without this gas burning does not take place—so in the bodies of plants and animals the complex substances derived from the elaboration of the food are gradually but very slowly burnt up. The energy thus set free enables animals to move about and, in certain cases, such as birds and mammals, to maintain a fairly high temperature. But this movement and this heat disappear as soon as the plant or animal dies. They are, in fact, attributes of *living* matter.

But the analogy between a fire burning and the oxidation of protoplasm is not an exact one. In a fire the law of "mass action" holds—the quicker oxygen be supplied, the faster burns the fire. In living tissue the rate of oxidation is not increased in accordance with the law of mass action. Only if the carbon dioxide be removed as rapidly as it is formed is this the case.

Every boy is perfectly aware that he is warm. When he goes to his chilly bed in a winter's night he knows that next morning the bed and the bedclothes will feel warm. He has been acting as a warming pan. Exercise which involves increased respiration makes a boy still warmer, and the fact that during a football match he is panting is an expression of the fact that he is using up his energy and becoming hotter during the course of the game. It is a common expression to hear that people put on thick clothing to keep the cold out. As a matter of fact they put on thick clothing to keep the heat in.

In the steam engine the combustion ultimately produces a certain amount of ashes, for the coal is not wholly consumed; and so would it be if instead of eating our food we dried it and burnt it—the result would be ashes. Carbon dioxide and water are being given off by all plants and animals, but the "ashes" of plants are more or less used over again and built up into new compounds in their bodies. In animals, however, the so-called "ashes," represented by urea and uric acid, etc., are excreted by the kidneys and the sweat-glands. Plants have no organs comparable with kidneys and sweat-glands.

Plants are built up more by carbohydrates than by compounds of nitrogen, and it is these substances which are mainly used up in plant respiration. In animals, however, the proteins with their nitrogen content are more dominant and the combustion of these necessitates special organs such as contractile vacuoles in the unicellular forms, flame-cells, the "kidneys" of many worms, and kidney-tubules which rid the animal body of its waste matter or "ashes." Most of the substances produced by the oxidation or burning of proteins are poisonous to all living cells. Therefore the animal throws

PLANT RESPIRATION

them out of its body; but the plant, being in every respect a more economical machine, changes their chemical nature and builds them up again into some sort of protein material. This means that the energy liberated by the breaking down of proteins is used in building them up again, so that far less energy is available in the plant for movement and for the production of heat. Plants move hardly at all, and the temperature even of a passion-flower or a tiger-lily varies not at all from that of the surrounding medium.

But an interesting fact has recently been noted. Just as our temperature rises when we are attacked by the bacteria which produce, say, diphtheria or enteric fever, so the temperature of a sick grape-fruit or an invalid orange rises by 1 to $1\frac{1}{2}$ per cent., or even 2 per cent. when the fruit is infected by a fungus, so long as it be alive. If the infection takes place in the dead fruit no increase of temperature occurs. These facts, which require further investigation, will doubtless greatly assist those in charge of the transport of fresh fruit, which has already attained enormous proportions, all over the world.

Plants have no respiratory organs. No special part of their body is set aside for the taking in of oxygen and the giving out of carbon dioxide. As we have seen, the plants, especially their leaves, are permeated by channels and spaces. The cells of the flowering plant do not fit close together; intercellular spaces exist where three or four cells meet. Into these channels and spaces the air penetrates and oxygen is taken up through the walls of the cells and through the same walls carbon dioxide passes from the cell into these intercellular spaces and then leaves the body of the plant. Plants, as it were, breathe all over and throughout the body. The period a plant can survive in the absence of oxygen varies. The higher plants, at any rate, do not live long in the absence of this gas. They succumb to the accumulation of such poisonous products as carbon dioxide. Even such inert things as seeds, if kept without oxygen, soon die.

Respiratory Organs in Animals

The lower multicellular animals, such as sponges and jellyfish, take in oxygen all over and throughout the body. The oxygen which these aquatic animals require is dissolved in the circumambient fluid. If removed from water they cannot live, for they cannot take up oxygen from the atmosphere. But this distinction must not be pushed too far, for in vertebrates the inner tissue of the lungs is moist. This moisture takes up the oxygen of the air and transmits it to the blood vessels. Air-breathing animals have about 21 per cent. of the atmosphere to draw on for oxygen, but a litre of average sea-water, under normal conditions, absorbs from the dry atmosphere only 6·44 c.c. of oxygen. To counterbalance the smaller proportion of oxygen available to aquatic animals, there is the extreme solubility of carbon dioxide in water. In consequence of this the carbon dioxide is removed very quickly from aquatic animals, especially from those that breathe partly through the skin, as the frog. This rapid removal of carbon dioxide may not appear of any great assistance to the animal, but carbon dioxide is a waste product, and, like the ashes in a furnace, will extinguish the flame if not removed. Carbon dioxide is, however, not entirely a useless product; it is of the greatest importance in controlling the rate of oxidation or combustion of the tissues. When carbon dioxide dissolves in water and in the body fluids, it forms a very weak acid; of course, the acid becomes weaker and weaker the less carbon dioxide there is in solution. Now supposing we take violent exercise, much carbon dioxide is produced which dissolves in the blood and renders it slightly more acid; this increase in acidity lowers the rate at which oxidation proceeds. On the rate at which oxidation proceeds depends our supply of energy, and if we can get rid of the carbon dioxide as fast as it is made we can keep our energy supply constant. Land animals breathing by lungs have to pant to rid themselves of carbon dioxide; water breathing animals need to make no such exertion, since in them the carbon dioxide diffuses into the surrounding medium so quickly that little or no carbon dioxide is found in the body fluids. The rate of oxidation being controlled by the carbon dioxide and not by the amount (tension) of oxygen, we see that aquatic animals lose nothing by having less oxygen available for breathing.

RESPIRATORY ORGANS

The essential feature of respiratory organs, the gill or the lungs, is the thin, moist membrane which separates on one hand the oxygen of the air or the water from the respiratory fluids of the body, generally "blood," and which permits of a free interchange of gases, oxygen passing in and carbon dioxide passing out. In many cases, such as leeches and earthworms, this membrane is the outer layers of the skin, and in certain leeches the minute capillary channels containing the blood actually lie between the neighbouring cells forming the epidermis or outermost layer of cells, a membrane only one cell thick. Many marine worms have processes which are regarded as gills. These may be extensions of the body wall and take the form of folded flaps or feathery plumules. Often a gill performs two functions. For instance, in the bivalve molluscs the bulk of the respiration is carried on in the inside of the thin mantle lining the shell, whilst the highly complicated so-called gills are in the main used for catching and conveying food to the mouth.

A gill or lung is obviously a delicate organ and is in most cases well protected: the lateral processes of the body which form the gills of many Crustacea are sheltered by a calcareous outgrowth of the skin and a special arrangement exists for pumping the water over these so-called *branchiae*. The gills of fishes are vascular processes on certain slits opening into the gullet. The oxygen-containing water is as a rule taken in by the mouth and expelled through these gill-slits, or it may be taken in and then expelled from the gill-slits without passing through the mouth. In nearly every case the gills are protected by a gill-cover or *operculum*. Some animals breathe by pumping water in and out of the hinder end of the alimentary canal. This is true, for instance, of certain Crustacea, of the sea-cucumbers, insect-larvae, and other aquatic animals.

The insect provides a remarkable example of efficiency in respiration. Here the oxygen is conveyed to and the carbon dioxide removed from the tissues, and even to and from the individual cells, direct, a much more efficient process than in other animals, where both the carbon dioxide and the oxygen have to pass into solution in the blood and then pass out again to the cell. The gain in energy with a practically unlimited supply of oxygen together with the rapid removal of carbon dioxide enables certain insects to contract and expand their muscles, and so to vibrate their wings, no less than 880 times a second.

In insects, centipedes, and spiders, instead of the blood being brought to a special organ, gill or lung, and there receiving its oxygen and then transmitting it to the innumerable cells of the body which it bathes, the air is taken straight to the tissues and cells of the insect or spider without the intervention of a respiratory fluid, "blood." Openings occur at the side of the body called *stigmata* or *spiracles*. These lead into ducts which, dividing and subdividing, spread in their final branches to every cell of the tissue. These tubes or *tracheae*, as they are called, are kept from collapsing by means of a spiral thickening, which resembles the metal wire in a garden hose or the coil-like thickening of the spiral-vessels in certain plants.

The direct supply of oxygen to each cell, without the cumbersome intervention of the blood, may to some extent explain the dominance of this class not only in variety of species but in number of individuals.

A very careful calculation of the number of species of animals in the different large groups, made in 1881, gave the following results:

Mammalia	2,300	Arachnida	8,070
Aves	11,000	Myriapoda	1,300
Reptilia and Batrachia	3,400	Insecta	220,150
Pisces	11,000	Echinodermata	1,843
Mollusca	33,000	Vermes	6,070
Bryozoa	120	Coelenterata	2,200
Crustacea	7,500	Porifera	400
		Protozoa	3,300

Here the total number of species of insects amounts to 220,150, whilst the remaining animals amount only to 91,503. I have recently had occasion to consult the authorities of the British Museum as to the number of known species. They estimate that mammals number 10,000; birds 16,000; reptiles and amphibia 9,000; fish 20,000; mollusca 60,000; crustacea 12,000 —probably an underestimate—whilst the number of insects is now put at 470,000, a little under half a million. These

RESPIRATION IN AMPHIBIOUS ANIMALS

figures strongly support the view that insects are the most dominant creatures on our earth.

The system connected with one spiracle is usually connected with that of the neighbouring spiracles by longitudinal tracheae, so that should one of these openings become blocked, air can still be distributed to those parts of the body which it primarily serves. In many insects, such as bees, there are large swellings or bladders in the tracheae which lighten the body and help it in flight.

It has been pointed out that those animals which are amphibious and are capable of living both in water and on land are animals which, as a rule, have well-protected gills. Certain crustacea and gastropod molluscs and a few fishes are able to breathe both in water and in the air. There is a certain little fish which skips about on the rocks above high water mark in the tropics, whose gills do not fill the gill chamber, and this chamber contains both water and air, and there are others, such as the climbing perch, which is said to be able to climb palm trees, having pouches from the gill chamber containing respirable air. When the water in which these fishes are living is insufficiently oxygenated they will come to the surface and swallow a bubble of air. Land-crabs found in many tropical islands have greatly enlarged cavities for their gills, and the walls of this cavity are richly supplied with vessels, and these walls act as lungs when the animal is out of the water. But these creatures always return to their old home to breed, and when they are reproducing the gills come into action again.

Dust and Sand

The fact that fine dust or sand in the air interferes with breathing is recognized by our Government in numerous Mining and Factory Acts, and fine dust and turbid water are equally bad for aquatic animals. During the great eruption of Vesuvius in April 1906 a very large number of marine animals were killed in the bay of Naples by the fine volcanic dust which fell on the waters and slowly sank. Even the sea-lilies, *Antedon*, were killed by the blocking of the openings of their *ambulacral pores*, which lead from the

106 RESPIRATION

exterior into the cavity of the water-vascular system. Another example of the same sort of trouble is that of the honey bee, which is infested at times with a certain small mite living in its tracheae. Its presence is a serious interference with the proper breathing of the bee, which is often killed by choking.

The amount of soot and dust in the air, owing to the fact that what laws we have are not enforced and are in themselves inadequate, is shown by the fact that, in the month of June alone, no fewer than 54 tons of dirt of various kinds were deposited from the air in the City of London, which covers an area of one square mile. Of this mass of dirt, approximately eighteen tons were soluble, and included sulphates, chlorine, and ammonia, and thirty-six tons were insoluble, and consisted of tar, soot, and grit.

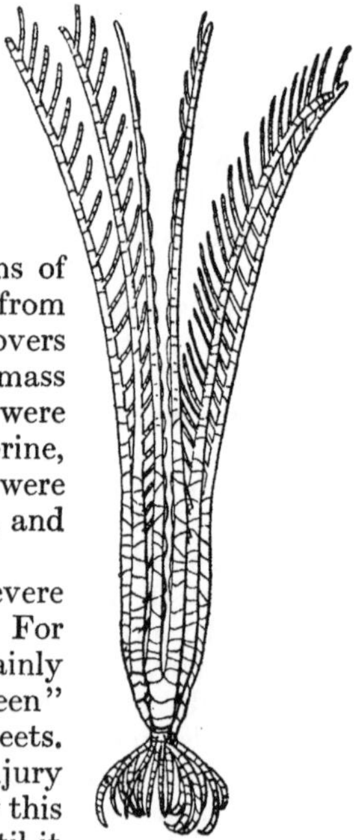

There could scarcely be a more severe indictment of our present methods. For the inhalation of grit and tar is certainly dangerous, while the "smoke screen" shuts out the sun-light from our streets. There can be no doubt that great injury is inflicted on animals and plants by this impurity and pollution, and that until it has been removed we shall fall far short of even a moderate standard of public health. Pure, sunny air, indeed, is the greatest need of city-dwellers at the present moment. The withholding of it, on any pretext, is unjustifiable.

FIG. 34. A Sea-lily, *Antedon acoela*. A young individual magnified 1½. After Carpenter.

Dove va il Sole non va il Medico.

HAEMOGLOBIN

We have seen that in insects, spiders, centipedes, etc., oxygen is conveyed direct to the living cells without any

HAEMOGLOBIN

intermediary; but in most animals it passes through the membrane of the gill or lung into a fluid. This fluid—in most cases blood—has a red substance in it which readily unites with the oxygen in the breathing organ and readily parts with it to a tissue or cell which is oxygen-hungry. The commonest of these substances, *haemoglobin*, always contains iron. It is found throughout the vertebrates and in some of the invertebrates, especially those that are at a disadvantage in the competition for oxygen.

As a rule haemoglobin is confined to certain cells called blood-corpuscles, which are in most vertebrates nucleated but in the Mammalia have no nucleus.

The distribution of haemoglobin is very peculiar. It occurs with one or two exceptions in special red corpuscles in the blood of all the Vertebrata, and it is found in corpuscles in the body fluid of certain bristle-worms, Chaetopoda. It also occurs in a species of *Solen* and also in *Arca*, its presence in the former being associated with the mollusc's very active life as a borer. And again it is not uncommonly found where strong muscular action is required: thus it occurs in the muscles of the odontophore of many GASTROPODS, snails, etc. Wherever increased facilities for respiration are required, there haemoglobin may turn up to assist the organs to carry on their work. Many other invertebrates have haemoglobin dissolved in the fluid of the blood, the *plasma*, and here it is not associated with special corpuscles. This is true of the earthworm and the medicinal leech and one other genus of leech. The same also is true of the larvae of a certain gnat, *Chironomus*, which lives in foul water, and in that of the gut of a common horse-fly, which lives in the alimentary canal of the horse, of certain primitive Crustacea including *Apus* and the ordinary water-flea, *Daphnia*, and of a water-snail, *Planorbis*.

Haemoglobin is diffused in the muscular tissues of mammals and of birds, and it is usually most apparent in very active animals like the hare, or in very active muscles like those used in flight by the black-cock. In both cases it gives a dark colour to the flesh. In reptiles it is confined to certain muscles, probably to those that are most active. The flickering dorsal fin of the sea-horse, *Hippocampus*, of our coast is tinged with haemoglobin, and so are the muscles of that active organ in the vertebrates, the heart. In many molluscs whose jaws are constantly working haemoglobin is found in the muscles of the pharynx, and curiously enough it is diffused throughout the nervous system of a certain bristle-worm, the sea-mouse *Aphrodite*, and in a certain Nemertine worm whose nervous system stands out brilliantly crimson in colour.

Haemoglobin, like other respiratory substances, combines loosely with oxygen, which it takes up from the air or the water and readily parts with to the tissues of the body. In the blood of man the red corpuscles exist in numbers which no one but a German Chancellor of the Exchequer could possibly cope with. There are 5,000,000 corpuscles in each cubic millimetre of blood, and there are 5,000,000 cubic millimetres of blood. So altogether we have inside us some 25,000,000,000,000 of the non-nucleated cells in the blood. They are circular and bi-concave, with a diameter of seven or eight μ. They vary slightly in size amongst the mammals, being smallest in the deer and largest in the elephant. In the camel they are bi-convex and oval—why, nobody has ever been able to explain. In no mammals do the red corpuscles have nuclei; but in all other vertebrates, fishes, amphibia, reptiles and birds, they are oval, bi-convex, and nucleated. They are in all these cases larger than those of mammals, and the biggest of all are found in certain tailed amphibia.

In some of the marine worms the respiratory pigment is known as *chlorocruorine*. It is green and gives a characteristic colour to its owners. But it is chemically altered by alcohol as is also haemoglobin. In the Crustacea haemoglobin is replaced by a fluid containing copper instead of iron. This is known as *haemocyanin*, and is colourless during life, but acquires a plumbago-like tint when exposed to the air. Anyone who has eaten a lobster will have seen this faint bluish colour in the larger vessels. The same oxygen-fixing agent in which copper largely takes the place of iron is common in many mollusca.

BLOOD

Besides the red corpuscles, blood also contains a considerable number of white corpuscles of five or six different kinds. In a healthy man the proportion of white corpuscles is about 1 to 500 or 600 red corpuscles, but the proportion is not constant, certain kinds increasing during a meal. Amongst other functions the white corpuscles, which resemble *Amoebae*, ingest or engulf foreign particles such as bacteria, and put them out of action. Those that perform this function are known as *phagocytes*. These cells are a great help to the body by fighting and destroying the germs of disease.

Like all other cells, blood corpuscles have a definite limit to their existence. After they have fulfilled their function, they waste away and die. A certain number, at any rate, undergo disintegration in the liver and in the spleen. To replace these, new corpuscles must be formed and this possibly takes place in the spleen, and certainly in the red marrow of the bones.

The total amount of blood in the body of mammals averages somewhere about one-sixteenth of the total body-weight. In a well grown man of 30 weighing 70 kilos there are about $5\frac{1}{2}$ kilos of blood. Many people think that it is the heat of the blood which keeps the body warm. In reality it is the other way about. The tissues of the body are continually but slowly burning, and as the blood courses through them it picks up their heat. Certain parts of the body are slightly warmer than others, but the circulation of the blood tends to keep the temperature fairly constant throughout the system. The normal average temperature of the human body is 98·4 F. degrees, but it varies a little during the course of the twenty-four hours, and it sinks during sleep. The temperature reaches its highest in certain forms of birds. For instance, the guinea-fowl has a temperature of 110° F., the pheasant 108·7° F., the greater titmouse 111·2° F., the swift 111·2° F. These high temperatures all indicate great bodily activity. Amongst mammals the duck-billed *Platypus* and the spiny anteater, *Echidna*, the most primitive members of the class, have a temperature which varies only from 2 to 5 degrees or even less than a degree above that of their external surroundings.

Cold-blooded animals include all the Invertebrata and such Vertebrata as the fishes, amphibia, and reptiles, but even these have as a rule a temperature a few degrees above that of the external surroundings. The viper has a temperature of 68° F. to 84° F., the frog of 58° F. to 63° F., the carp 69° F. and leeches 57° F. Although a single bee becomes numb and inert at the temperature of a summer night, the vast numbers of individuals in a hive produce a considerable amount of heat, enough to keep the bees active during the winter. In December at a time when the external air was only 1·57° F. the temperature in the interior of a hive has been found to be 22·8° F.; and this is indeed necessary, as the larvae, pupae, and eggs all die in a temperature below 17° F. It is also necessary in order to keep the wax soft and malleable.

The tissues of an animal are constantly discharging carbon dioxide into the blood and picking up oxygen from it. Both the corpuscles and the plasma contain carbon dioxide, and blood contains about the same proportion of carbon dioxide as water does when it has been shaken up in the presence of that gas. In mammals blood is driven from the right side of the heart to the lungs. Here it frees itself of its carbon dioxide and picks up oxygen. The arterial blood, as it is then called, returns to the left side of the heart and is driven round the body, where it readily yields up oxygen to the oxygen-hungry cells, and picks up carbon dioxide before regaining the right side of the heart. It passes from the arteries through the capillaries to the veins and so back to the heart.

The number of capillaries is enormous, though that of those that are open at the same time in a muscle varies according to its state of activity. The section of an ordinary pin is about half a square millimetre, yet in a square millimetre of the muscles of a horse there are 3150 capillaries. In smaller animals such as the guinea-pig the number per square millimetre is even higher. This means that a very large surface of blood is available for interchange to take place with the cells of the tissue. Krogh has made the following calculation: "Supposing a man's muscles to weigh 50 kilograms and his capillaries to number 2000 per square millimetre, the total length of all these tubes put together must be something like

LUNGS IN MAN

100,000 kilometres or two and a half times round the globe, and their total surface 6300 square metres."

In an ordinary healthy adult the heart contracts from 70 to 80 times a minute, but in infants and youths it is quicker, and in old men much slower. In fact, the number of pulsations per minute greatly diminishes throughout life. It has been said that the work of the heart in a day is equivalent to that done by an able-bodied labourer working hard[1] for two hours. Its muscles are the most hard-worked of the body: night and day they have no rest.

In man the lungs are not very big. Perhaps on an average they equal in bulk, when collapsed, both closed fists. But they expose an enormous area to the air which they breathe in, and to the blood which circulates in their tissues. The object of all breathing organs is to afford as much surface as possible so as to facilitate the interchanges of gas. A lung consists of an enormous number of little pockets or *alveoli*, of which there are said to be in the human lungs some 725,000,000. Their total surface has been calculated at from 90 to 95 square metres (107·6 to 113·5 square yards), an area which would carpet a large room of 30 feet by 36 feet. Since a suit of clothes for an average man takes 3 to $3\frac{1}{2}$ square yards, the internal surface of the two lungs is equal to that of some thirty suits. Normally the total quantity of air which passes in and out of the lungs of an adult at rest in 24 hours varies from 400,000 to 680,000 cubic inches.

A man can live some eight weeks without food and more than a week without water, but only for a few minutes without oxygen; and this is true of all warm-blooded vertebrates. Lack of air induces death much quicker than lack of drink or food. Diving birds as a rule only remain under water for something less than a minute, and the most expert pearl-fishery divers in the East must return to the surface after two or three minutes. On the other hand, so highly adapted are whales and other Cetacea to their watery surroundings that sperm whales have been known to remain beneath the surface from an hour to an hour and twenty minutes; and many species remain under water from twenty minutes to an hour.

[1] This estimate is a pre-war one.

RESPIRATION

ANAEROBES

We now come to the exceptions mentioned in the first line of this chapter. When I was a student we used to be taught that all plants and animals took in oxygen and breathed out carbon dioxide. Further investigation, however, has shown that this statement is not true of all plants or animals. Many of the bacteria, like the vast majority of other living organisms, can only live in the presence of free oxygen. These organisms are termed *aerobic*. Many others, and such organisms as the yeast plant, can live without free oxygen for a considerable time, attaining the energy for their growth by breaking up sugar into alcohol and carbon dioxide. But after a time they require oxygen. Others can live entirely without oxygen, whose presence is often fatal to them, obtaining their energy by breaking down complex compounds. They are called *anaerobic*.

After Lavoisier's classical work on oxidation the idea of the impossibility of life without oxygen dominated the minds of all scientific men for many years.

Spallanzani, contemporary with Lavoisier, proved however, experimentally, that small infusorial animalcules and snails live and produce carbon dioxide when the supply of oxygen is entirely cut off. This very important investigation of Spallanzani remained unnoticed until the year 1861, when Pasteur published his first monumental note on "Animalcules infusoires vivant sans gaz oxygène libre et déterminant des fermentations." In this note Pasteur introduced for the first time the terms *aerobic* and *anaerobic*. Since Pasteur much work has been done on this subject and we know now that while some micro-organisms can live temporarily without oxygen, for others which are permanently anaerobic even a small oxygen tension is a deadly poison.

In 1875 Pflüger showed that frogs can be kept for twenty-four hours or more in pure nitrogen and that during this time they produce a considerable amount of carbon dioxide.

The presence of free oxygen is poisonous to certain anaerobic organisms: it kills them. The yeast cells on the other hand cannot live for an indefinite time without oxygen, for the

ANAEROBIC ORGANISMS

alcohol they produce ultimately poisons them, and this is true of many anaerobic bacteria. It had been known for some time that the bacterium causing lock-jaw, *tetanus*, was one of these anaerobes, and in the distribution of such organisms the soil, especially if highly manured, plays a great part. There is no free oxygen, or practically none, in the alimentary canal, and many of the bacteria that live there are anaerobic. Many anaerobes are extraordinarily poisonous; and during the war the extensive character of the wounds caused by shell splinters or contaminated by the earth gave ample opportunity for infection. Gas-gangrene, for instance, was caused by one of these organisms, *Bacillus welchii*, and during the first months of the war it was very difficult to deal with. Another one is *Bacillus botulinus*, which occurs from time to time in potted meats and vegetables, and as we have recently seen, may be the cause of sporadic outbreaks of very fatal disease.

But the power to exist without oxygen is by no means confined to the lower animals. We have said above that there is practically no free oxygen in the intestine, and yet many worm-like organisms live therein. Tapeworms which attain a length of 36 metres, and round-worms which may attain a length of from 16 to 45 centimetres and a breadth of 6 millimetres, live in the intestine in an atmosphere devoid of oxygen. I have myself taken from the air-bladder of a rainbow trout, caught in a stream on Lord Knutsford's estate near Royston, a number of perfectly healthy round-worms. An analysis of the air in the trout's air-bladder, immediately after death, gave the following results:

Carbon dioxide 1·5 per cent.

Oxygen 0·0 per cent.

Nitrogen 98·5 per cent.

There was not the faintest trace of oxygen, and yet the round-worms flourished and bred. There is still very much to be learned about the metabolism of these highly developed creatures which live in an oxygen-free atmosphere. It has been shown in the case of the two round-worms, *Ascaris mystax* and *Ascaris megalocephala*, which live in the intestines of the cat or dog and of the horse respectively, where the oxygen tension is almost nil, that they do not use oxygen at all, in spite of which they produce considerable quantities of carbon dioxide. Even when oxygen is offered them they are unable to make use of it. On analysing such worms it has been found that one-third of their dry substance is composed of pure glycogen, and their energy—which of course is not much, as they move but little and derive their temperature from their host—is obtained by the breaking down of this glycogen into carbon dioxide and certain fatty acids. Somewhat similar processes take place in earthworms and in certain leeches, *Clepsina* and *Nephelis*, when kept in an atmosphere of pure nitrogen. Leeches will live in such an atmosphere from two to six days.

CHAPTER XI

MOVEMENT

MOVEMENT OF PLANTS—AMOEBOID MOVEMENT—MOVEMENT OF ANIMALS—FLIGHT

This is the law of the beast, and of the fowl, and of every living creature that moveth in the waters, and of every creature that creepeth upon the earth. Leviticus xi. 46.

Movement of Plants

ANIMALS, with the exception of the few that are fixed, *sessile*, such as coral-polyps, oysters, etc., can and do move

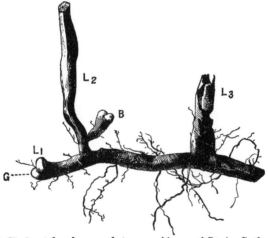

Fig. 35. Horizontal underground stem or rhizome of *Pteris*. G, the growing point; L_1, a developing leaf; L_2, the leaf of the current year; L_3, a decayed leaf of the previous year. The rhizome bears roots; from L_2 a young rhizome B is growing. (Darwin's *Elements of Botany*.)

from place to place; plants, as a rule, cannot. The more primitive unicellular plants living in water are of course moved about by the streams and currents, but this is no movement of their own. Certain of them, however, propel their bodies through the water by means of the lashing of a long, whip-like flagellum. Very often there are two flagella, and sometimes there are cilia. Then, again, the antherozoid of plants—the male cell, corresponding with the animal spermatozoon—which fertilizes the egg of the plant, is motile. For instance, in a fern the antherozoid swims through water to make its way to the egg or ovum, and this is true

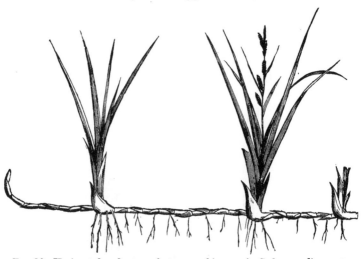

FIG. 36. Horizontal underground stem, or rhizome of a Sedge, sending roots downwards and leaves upwards. From Le Maout and Decaisne.

of many plants, even as high as the sacred Chinese maidenhair tree, *Gingko*. The bracken-fern, *Pteris*, does, however, move. Its creeping underground stem, or rhizome, is always pushing forward in the soil. The front end grows and the hind end decays. The leaves or fronds are borne near the latter end, they die down during the autumn, and the next year's fronds will be borne an inch or two further along the rachis, nearer to the front end, to the growing-point. Thus the whole fern, slowly, very slowly, moves forward. Sedges and some other plants also have underground stems which

MOVEMENT IN PLANTS

creep through the soil. But the vast majority of plants, trees and shrubs and flowering plants, and mosses and mushrooms and seaweeds are rooted in the soil or to some base, and their food flows round them: they have no need to go after it, nor after the oxygen which they respire. But their immobility has its dangers: they cannot move away from a vitiated atmosphere, or from the neighbourhood of danger, or from anything that bores them.

Although few plants move as a whole and these almost entirely water-plants, their parts move. We have seen above that the protoplasm in the cells is in a constant state of

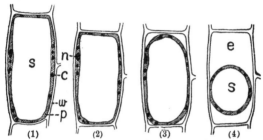

FIG. 37. Stages in the collapse of a cell when immersed in a strong solution: in (1) it is fully turgid, with the cell-walls distended; in (2) the cell-wall is no longer stretched, and the cell is therefore flaccid but the protoplasm is still in contact with the cell-wall; in (3) plasmolysis has begun, and in (4) the protoplasm has rounded off and only remains in contact with a small part of the wall. *w*, cell-wall; *s*, cell-sap; *p*, protoplasm; *n*, nucleus; *c*, chloroplast; *e*, solution which has passed through the wall of the cell. After De Vries, modified.

movement. It circulates and flows hither and thither. But although in all higher plants the whole organism does not move—is not in fact locomotive—parts of the whole do move. This they generally do by their tissues stretching and contracting, and this is brought about by alterations in the amount of fluid in the cells. Cells may be *flaccid* or *turgid*, limp or full of water and stiff. Alternations in the turgidity of the neighbouring cells produce movement. An actively growing stem usually advances in an irregular spiral due to the growth of one side of the growing end being more rapid than the other. Then again, the so-called telegraph-plant of India has leaves which, apparently for no reason, slowly twist round so that their apex describes a circle in the course of one and a half to two minutes. Another movement of the plant body is shown by twining stems. The hop, the scarlet runner bean and the *convolvulus* revolve round sticks or other supports.

Fig. 38. *A*, Hop twining with the sun; *B*, *Convolvulus* twining against the sun. After Payer.

Tendrils are another example of movement. They are, as a rule, whip-like organs with a hooked tip. They revolve in a spiral way, and when the tips come into contact with a support the tendrils will wrap themselves round it. Morphologically a tendril is usually a part of a leaf, but this is by no means always true. But there are other movements which are due to external conditions.

TROPISMS

Many of the movements of plants and parts of plants are stimulated by light, heat, gravity, electricity, etc. The plant responds to these external conditions. Movements due to the stimulation of external force are called *tropic*. A rooted plant is so constituted that its root grows downwards and its stem upwards, and if the root be placed horizontally it will in time bend downwards. The root is positively *geotropic*—seeking the earth—whereas the stalk is negatively geotropic—avoiding the earth. These movements are, of course, of value to the plant inasmuch as they help to keep its various organs in the right places for carrying on their functions.

Light plays a large part in the movements of the various organs of the plant. Like the ex-Kaiser, green plants seek "a place in the sun." The majority of plants are *heliotropic*; thus the ordinary green leaf places itself so that it can receive the largest number of rays of light. Roots seldom or never contain chlorophyll and are negatively heliotropic. Certain chemical substances which pass from the cells of the plant into the surrounding media induce movements in a given direction, for instance, the antherozoids of the bracken-fern are said to be attracted to the egg-cell by the excretion of malic acid.

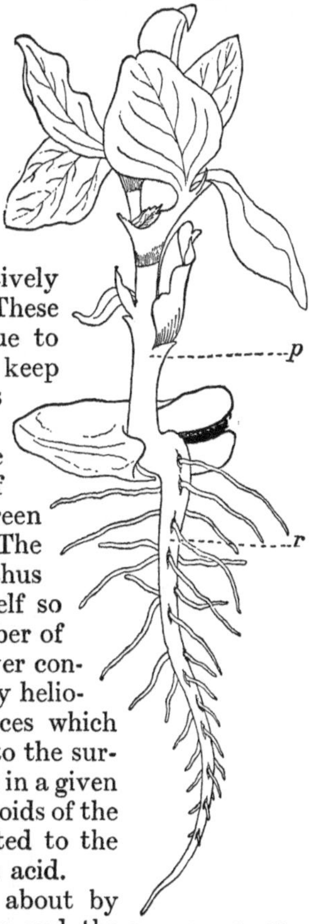

Other movements are brought about by touch. The leaves of the sundew and the Venus' fly-trap close up and entomb any insect which rests on their surface. The stamens of the barberry and of some orchids also move when touched, and then again we have the well-known sleep movements; the leaves of the

FIG. 39. Seedling of the Bean, *Vicia Faba*. p, the stalk; r, the root. (Darwin's *Elements of Botany*.)

120   MOVEMENT

clover, the acacia and many other plants adopt different positions during the day and during the night. They show that rhythm which exists throughout all living creatures. The sensitive plant, *Mimosa pudica*, is, as its name implies,

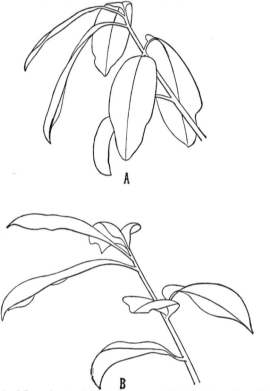

FIG. 40. A Laurel twig, *A* in the frozen, and *B* in the thawed condition. From Darwin and Acton.

uncommonly sensitive. At the base of the leaf-stalk and at the base of the secondary leaf-stalks, there are swellings. These are composed of turgid cells, and any stimulus such as touch, changes of light, temperature, or chemical stimulus will act on those swellings: the leaf-stalks droop, the leaflets fold together, and the whole thing hangs down. This plant also shows that a stimulus applied at one part is transferred to other distant parts, presumably through the protoplasmic threads which connect up the contents of the cells.

Finally we have movements which are brought about by the contiguity of thick-walled, stiff cells and thin-walled limp cells. Should water be withdrawn from the thin-walled cells they shrink, whereas the thick-walled cells retain their water. This shrinkage causes the leaf to curl up, but it will uncurl when water is supplied to the thin-walled cells.

Occasionally parts of the plant explode, and this is due to certain cells being extremely weak and feeble, whereas others become extremely tense and turgid. A slight increase of the turgidity will rupture the weaker cells. These explosive movements are often connected with a dispersal of seeds or spores; the squirting cucumber is an example of the former, and in fact many fruits exhibit this action. The common geranium and the violet jerk out their seeds, and at times the explosion is accompanied by an audible sound. Plants are as a rule mute: no one expects a mushroom or a beech tree to burst into song. They have the great gift of silence, and when they do make a noise it is a momentary pop and is seldom repeated.

Amoeboid Movement

Amoeboid movement is to be seen in almost every group of the animal kingdom. In the protozoa many animals other than those classed with *Amoeba* itself are amoeboid at certain stages of their career. In the higher animals there are many cases of this form of movement; the cells lining the alimentary canal of the *Coelenterata* and of the *Turbellaria* are often amoeboid; the nerve-fibre or *axon* of a developing nerve reaches its destination by amoeboid movement; almost every group of animals have *amoebocytes* or wandering cells which exhibit amoeboid activity.

How does amoeboid activity take place? All that we can observe is that the protoplasm of *Amoeba* is capable of great changes of shape although there is no visible internal contractile mechanism such as the fibrillar structure observed in many muscles, etc. It was formerly thought that amoeboid movement took place owing to local differences of surface-tension at the surface of the *Amoeba*. If the surface-tension was weakened at one spot the protoplasm would bulge out there just as a rubber balloon bulges at a place where the rubber is weakened. But the problem is not so simple. Surface-tension alone is probably not powerful enough. The mechanism of amoeboid movement is still a mystery, but it is becoming more and more probable that it is bound up with the colloidal nature of protoplasm which turns readily from a more solid *gel* state to a more liquid *sol* state and *vice versa*.

I have not been able to find any record of the rate of progression of an *Amoeba*, so I asked a friend of mine to determine its average speed. He timed four separate *Amoebae*, very large *Amoebae*, for they measured when fully extended 0·5 mm. Altogether he made ten different observations, and he recorded the time the *Amoebae* took to pass over a measured distance of 0·2 mm. The fastest *Amoeba* covered the distance in 60 seconds: the slowest took 80 seconds. On the whole they averaged 70 seconds. At this rate their speed amounted to 10·3 mm. an hour. Unlike Browning's hero who "never turned his back, but marched breast forward," they sauntered leisurely along, pausing here and there, sometimes to throw out a pseudopodium on one side or the other, and then withdrawing it they moved forward again, like a man dawdling down a village street, stopping a moment to look at some object in a shop window, and a few minutes later to exchange the time of day with a neighbour. Their progress was discontinuous, so that the actual rate of their fastest movement is quicker than the above figures indicate. On a perfectly clean slide, however, they do not dawdle.

Movement of Animals

The great majority of animals can and do move from place to place; and even when they are fixed or, as it is termed, *sessile*, their larvae are capable of very active movement. Like all young animals they wish to scatter and see the world. They swim away from their mother and seek "fresh woods and pastures new."

MOVEMENT OF INVERTEBRATES

Perhaps the simplest form of movement is that already described in the *Amoeba*, where a lobe or pseudopodium is thrust out and the body of the animal slowly flows after it. It is, of course, an incredibly slow means of progression, and the advance made in twenty-four hours is well under one foot.

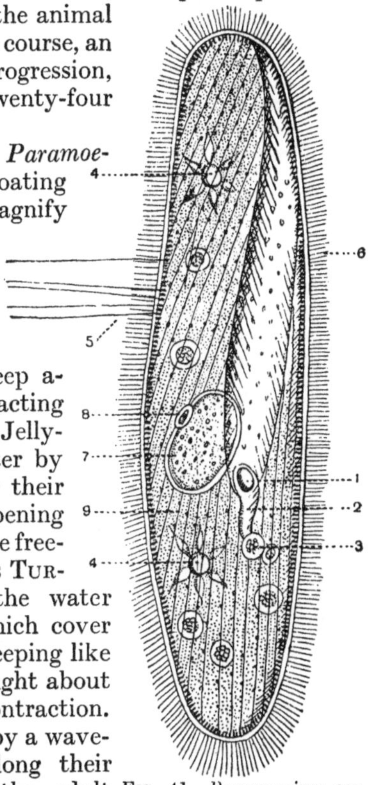

The slipper animalcule, *Paramoecium*, swims by means of a coating of cilia, and if you could magnify a *Paramoecium* to the size of a pony, it would move about as fast as a pony galloping. As a rule sea-anemones are fairly stationary, though they sometimes creep about by expanding or contracting their muscular foot or base. Jellyfish swim through the water by extending and compressing their hollow umbrella, like the opening and shutting of a parasol. The free-living flat-worms known as TURBELLARIA move through the water both by means of cilia, which cover their skin, *gliding*, or by creeping like a slug. The creeping is brought about by waves of muscular contraction. They will also swim freely by a wave-like contraction passing along their flattened body. Although the adult flukes, TREMATODA, and tapeworms, CESTODA, are parasites living with not much movement in the body of some host, usually a vertebrate, their larvae are capable of considerable movement, and indeed they need to be or they would never find new hosts. Roundworms, which again are largely internal

FIG. 41. *Paramoecium caudatum*, magnified about 250. After Bütschli. 1. Mouth at bottom of groove. 2. Gullet. 3. Food vacuole just being formed. 4. Contractile vacuoles. 5. Trichocysts or stinging hairs which have exploded: the unexploded ones line the cuticle. 6. Cilia. 7. Large nucleus. 8. Small nucleus. 9. Contractile fibrils.

parasites, move by the wriggling of their body, and their advance in the right direction is aided by the fact that their cuticle is ringed. A longitudinal section of the skin resembles

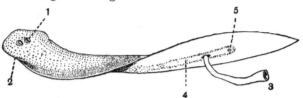

FIG. 42. *Planaria polychroa* swimming, magnified about 4. 1. Eye. 2. Ciliated slit at side of head. 3. Mouth of proboscis. 4. Outline of the pharynx sheath into which the pharynx can be withdrawn. 5. Reproductive pore.

the teeth of a saw, and the projecting ridges enable them to move forward as their body sways from side to side.

The great group of molluscs is so diverse that it presents a great variety of modes of movement. Many of them swim.

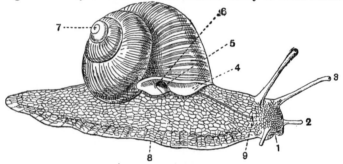

FIG. 43. *Helix pomatia*, the edible snail. Side view of shell and animal expanded and creeping and gliding along. From Hatschek and Cori, old types. 1. Mouth. 2. Anterior tentacles. 3. Eye tentacles. 4. Edge of mantle. 5. Respiratory pore. 6. Anus. 7. Apex of shell. 8. Foot. 9. Reproductive aperture.

Pecten, the shell of which used to be worn by pilgrims, swims by means of a curious jerky motion, alternately opening and shutting the valves of its shell. Some molluscs move by gliding. The snail, for instance, secretes a mucus from a gland which traverses the foot and opens just under the mouth, and the cilia which cover the lower surface of the snail's foot "row" in this mucus and the whole animal slowly glides

MOVEMENT OF INVERTEBRATES

along, leaving behind it a glairy deposit which can often be seen on the windows of greenhouses. Slugs move in somewhat the same way, and will travel at the rate of a mile in eight days, whereas the ordinary snail takes almost twice as long to cover the same distance. But the slugs have another method of progression, for they have muscular waves or contractions passing along the foot backwards at the rate of thirty to fifty a minute. Many fresh-water snails glide in the same way over the surface of water plants, and sometimes on the under side of the surface-film of the water. Some of these fresh-water molluscs have a curious habit of throwing out strands of mucus which act like a rope up which they can climb. Similar strands of mucus, which harden into threads, are emitted by certain slugs which descend quickly from plants to the ground and return up the same rope. Cuttlefish have a peculiar means of movement. By contracting a part of their body called the mantle, they squirt out a quantity of water which propels them backward, for a cuttlefish generally moves with its hinder end first.

The starfish and sea-urchin move by means of their tube-feet, "fingers," which, when they stretch them out, adhere to the substratum; then they contract and thus pull the body after them.

> And if "you" doubt the tale I tell,
> Steer through the South Pacific swell;
> Go where the branching coral hives
> Unending strife of endless lives,
> Where, leagued about the 'wildered boat,
> The rainbow jellies fill and float;
> And, lilting where the laver[1] lingers,
> The starfish trips on all her fingers;
> Where, 'neath his myriad spines ashock,
> The sea-egg ripples down the rock;
> An orange wonder daily guessed,
> From darkness where the cuttles rest,
> Moored o'er the darker deeps that hide
> The blind white sea-snake and his bride
> Who, drowsing, nose the long-lost ships
> Let down through darkness to their lips.
>
> KIPLING, *Many Inventions.*

The sea-urchin also helps itself along by means of its five calcareous teeth.

[1] Laver: an edible sea-weed.

Most of the jointed animals, ARTHROPODA, run or crawl about on their legs. Their legs move in a definite order. Water-worms and leeches generally move by the undulations of their bodies, swimming gracefully through the water. The latter also loop like certain caterpillars, GEOMETRIDAE.

FIG. 44. Underview of a Starfish, *Echinaster sentus*, with tube-feet "fingers" extended. Magnified about 1. From Agassiz.

Earth-worms push their way through the soil by means of small stiffened chitinous hairs called *chaetae*, which are arranged in pairs, so many to each segment. They can even climb up almost perpendicular surfaces, provided that the surface be rough.

A great many animals swim. Fishes do that by the waggling of their body, especially their tail, and the fins take a com-

MOVEMENT OF INSECTS

paratively small part in locomotion and act mainly as balancers. Crocodiles also swim by the lashing of their tail, but turtles and tortoises use their limbs.

Many animals move by jumping. A group of lowly insects called springtails have a mechanism very much like that which enables toy wooden frogs to jump from the ground. The extraordinary predominance of insects is shown by the fact that the British Museum contains 3,500,000 specimens of insects and only 1,500,000 other invertebrates; the latter cover the enormous classes of protozoa, coelenterates, worms of all sorts, crustacea, spiders, millipedes, mollusca, star-fish, and many other smaller groups. Not only are insects by far the most numerous animals both in number of species and of specimens, but they are relatively the strongest.

The muscles of the wings of insects contract far more rapidly than any other muscles known amongst all animals. It is held by a good many authorities that the buzzing of the mosquito is due to its wings, and certain experiments have been made upon the note emitted by both sexes. It was found that the males gave a higher-pitched note than the females, and that the note was higher in both sexes when they had fed; the greater the meal, the higher the note. This I have also noticed at "bump suppers" and City dinners. Of four unfed females three gave notes within a quarter of a tone of 264 (*i.e.* of 240 to 270 vibrations per second), the fourth female gave an abnormally low note of about 175 vibrations. Four other females were arranged in the order of the distension of the abdomen by food, the last being largely distended; these gave notes corresponding roughly to 264—281—297—317 vibrations or according to the musical scale, the notes:

Three unfed males gave exactly the same note, viz. corresponding to 880 vibrations  Immediately after feeding one gave the note A♯, another which had fed well B♮.

In insects the vibrations of the wings are untiring, and they maintain a uniform and amazing rapidity. A fly's wing vibrates 330 times a second, that of a bee 190 and that of a wasp 110. The strength of insects is also astounding. A bee can and does lift and carry 25·3 times its own body-weight, whereas a man can only lift and carry about the weight of his own body.

One of the best known jumping animals is the common flea. It has extremely powerful legs borne on projections sticking out from the sides of the body. Fleas do not jump anything like the height or length that you think they do when you are trying to catch them. The present record is held, like so many records, in California, by a Californian flea whose high jump amounted to $7\frac{3}{4}$ inches and long jump to 13 inches. I have recently had nine fleas weighed in a chemical balance by an expert. The average of the nine came to 0·38 mgm. An average man weighs some 70 kgm. Hence if a man had the same leaping power as a flea he could cover 36,800 miles in a horizontal jump, $1\frac{1}{2}$ times round the world. Vertically he could leap 21,900 miles. He could leap to the moon in about ten jumps. But of course this could never really happen, for the velocity of the athlete would be so terrific that he would burst into flame and disappear like a meteor in a cloud of glory. Kangaroos and jerboas are further instances of animals that progress by leaps and bounds.

Flight

When we come to the air we find two dominant groups which have captured that realm of Nature, the insects and birds. But by no means all insects fly. Neither the flea nor the louse nor the bed-bug has wings; yet all of them have an uncanny power of getting where they want.

Many other insects are devoid of wings, and either run about on their legs or, as in the case of the spring-tails, proceed by jumping. But the great majority of insects have gauzy wings which enable them to make considerable flights. Owing to its importance as a disease carrier the flight of the common house-fly has been carefully studied. Marked flies have been

FLIGHT

taken within 48 hours at distances varying from 300 yards to a mile; but the direction of the wind undoubtedly has a great influence on the distance which they travel. Mosquitos as a rule do not fly very far, but are frequently carried many miles by a strong wind. On a still day they will not cover more than a few hundred yards.

The flight of birds is of three kinds. Firstly, by the active stroke of the wings. Secondly, by gliding or skimming, supported on the outstretched wings which do not flap up or down; to glide or skim a bird must have a certain velocity, acquired either by previous strokes of the wings or by slanting down from a height like a parachute or by commencing flight in a wind of some velocity. Thirdly, by sailing or soaring by means of extended wings. This can only take place in a high wind, and is only possible with certain birds. It differs from gliding by the fact that the bird does not lose either in velocity or in vertical position as a result of the resistance of the air to the bird's passage. The muscles which move the wings up or down are very bulky and average about one-sixth of the weight of the entire bird; but in strong fliers such as the house-pigeon the proportion is 45 per cent. of the total body-weight. Many birds fly immense distances. The swift, for instance, which is but a short summer visitor in Great Britain and begins to disappear early in August, passes its winter in Central Africa. These birds have greatly developed salivary glands, the secretion of which forms the transparent nests so beloved by the Chinese for making soup. Our swallows migrate and winter far beyond the equator in Africa. The curious thing is that the young leave England some weeks before their parents, and without experience and without guidance find their way to their winter home. The Asiatic swallow will get as far as India or Burma and even farther south, occasionally as far as Australia, whilst American species winter in Southern Brazil. The American golden plover breeds in the extreme north of Canada and winters in Southern Brazil and the Argentina. But these great distances are equalled or almost equalled by the larva of the eel, which makes its way unguided and untaught from somewhere between Bermuda and the West Indies to the rivers of

130 MOVEMENT

Western Europe. As a rule the larger members of a family fly faster than the smaller. Thus a grouse surpasses a partridge

FIG. 43. Map showing the range of the American Golden Plover, with its known migration route. (By permission of the Bureau of Biological Survey, Washington, U.S.A.)

in speed and a black-cock a grouse. The swan flies faster than the duck. Under ordinary conditions game-birds fly

EFFICIENCY OF MUSCLES

about 40 miles an hour but when pressed can fly faster. Swifts which have been carefully timed cover a distance of 171·4 to 200 miles an hour.

In the higher animals locomotion is brought about by the contraction of muscles. The muscle becomes shorter and thicker. It will contract till it is from 65 to 85 per cent. shorter than before. But its volume scarcely changes.

From the point of view of doing work, the muscle is a very efficient instrument. An ordinary locomotive loses about 96 per cent. of its available energy in the form of heat, and only 4 per cent. is left to represent the work done. Even in the very best triple expansion steam-engines the work done rises but to 12·5 per cent. of the total available energy. In the vertebrates, muscles form by far the larger bulk of all tissues, some 45 per cent. of the whole body-weight. Muscles are the chief source of the heat of the body, and the heat of the muscles is increased when they contract. This heat is furthermore not wasted as it is in an engine. It keeps the body at a temperature necessary for life and health. In man from 20 to 28 per cent. of the energy liberated during the contraction of the muscles appears as work—five to seven times as much as in a locomotive. Muscles absorb more oxygen and release more carbon dioxide than any other tissue. A hundred grammes of muscle will take up 5·68 cc. of oxygen and give out 5·08 cc. of carbon dioxide per minute, whereas the same amount of the next most active organ, the brain, takes up but 4·58 cc. of oxygen and releases 4·28 cc. of carbon dioxide. Certain drugs will put the action of muscles out of control, and when that is done the temperature of the body drops.

Normal muscular *contraction* is an anaerobic process—as opposed to *relaxation*, which involves the oxidation of lactic acid (which has been produced from the glycogen). It is conceivable therefore that if proper mechanisms, *i.e.* not requiring oxidation, were provided for the removal of the lactic acid which provokes a muscle to contract, movement could occur indefinitely in the absence of oxygen. The need for oxygen in higher animals seems to be concerned more intimately with the nervous processes.

When processes of oxidation are diminished, as by potassium cyanide, a muscle is still able to contract, but lactic acid accumulates more rapidly than in normal muscle. Apparently, therefore, in normal muscle, part of the lactic acid is utilized (oxidized) to reconvert the remainder (about five-sixths) into glycogen, which greatly increases the efficiency of muscle. In severe exercise the lactic acid escapes into the blood and is excreted in the urine. If all of it were disposed of in this way, contraction—though inefficient—would be entirely anaerobic.

CHAPTER XII

RHYTHM

RHYTHM IN PARTS OF CELLS—RHYTHM IN CELLS—RHYTHM
IN TISSUES—RHYTHM IN ORGANS—RHYTHM IN ORGANISMS
—RHYTHM IN COMMUNITIES

> So do flux and reflux—the rhythm of change—alternate and
> persist in everything under the sky.
> THOMAS HARDY, *Tess of the D'Urbervilles.*

WE have said in Chapter I that one of the characteristics of living matter is rhythm: in fact, the whole of Nature shows certain rhythms. Summer and winter, seed-time and harvest, night and day, follow one another at regular intervals. In the sonorous words of Addison:

> Th' unwearied Sun from day to day
> Does his Creator's power display.
>
> . . . .
>
> Soon as the evening shades prevail
> The Moon takes up the wondrous tale,
> And nightly to the listening earth
> Repeats the story of her birth.

These external influences have their effect on both plants and animals, and are responsible for many of their rhythms. But apart from external stimuli there is an innate rhythm in living matter.

RHYTHM IN PARTS OF CELLS

Parts of cells may have rhythm. The contractile vacuoles of the *Infusoria* contract at intervals which are sometimes regular and sometimes irregular. There seems to be a great deal of variation in the rate of contraction of the vacuoles in different individuals, in the same individual at different times, and in the two vacuoles in the same individual.

The times between successive contractions of the vacuoles of *Paramoecium* were observed and found to vary from 15 secs. to 40 secs. *with an average of 27·5 secs.* In one case where the two vacuoles in the same individual were observed it was found that the posterior vacuole varied from 15 secs. to 30 secs. with an average of 22·5 secs., whereas the anterior vacuole contracted fairly constantly about every 35 secs. The rate of pulsation varies with the amount of salt in the surrounding water.

The protoplasm circulating in the living plant-cell is again rhythmic. It passes up the sides and down the middle in a definite, cyclic flow. If it bears the nucleus with it, the nucleus is at one moment of time at one end of the cell and a little later at the other end. It passes from end to end rhythmically.

Amongst the most prominent examples of rhythm in parts of cells are the cilia. They move independently of nervous impulse. The stimulus that induces them to beat comes from the protoplasm of the cell, and if this be completely removed the cilia cease to act. The cilia of the infusoria bustling about in a puddle, or those of the cells lining our breathing tubes, beat in unison and at a definite rate for each individual. Thus the beat of the first cilium is followed by the beat of the second, third, and so on. No single cilium ever contracts out of order, and it does not make its appropriate movement till the preceding cilium of the row has moved. It bends just after its neighbour has stopped bending. The first one, as it were, gives the signal to the others. Like the "stroke" in an "eight," it sets the pace. If it ceases to beat, all the others cease to beat; if it starts beating again, the others follow suit. The work they perform is said to be much less than that of muscular movement, and it has been determined that a *Paramoecium* 0·25 mm. in length would be able to lift about nine times the weight of its own body by its ciliary action.

The flagella of antherozoids or of spermatozoa also beat rhythmically. Their vibrations can be broken up and made irregular by external factors, but in normal health they beat in unison.

Rhythm in Cells

Rhythm in cells is perhaps best shown amongst the unicellular plants and animals. They are born; they grow up; they reproduce; and sooner or later the great majority of them die. Thus the malarial organism which infests the red corpuscles of man passes through its life-cycle in two or three days according to the species. The moment when it bursts out from the red blood corpuscle is the time when the malarial patient suffers his highest fever, and this occurs rhythmically every two or three days.

Many of the pigment cells which occur in animals are rhythmic. They are influenced by light and darkness, day and night. The pigment-cells of the chameleon-shrimp, *Hippolyte*, by contracting and expanding are able to vary the colour of the crustacean so that it resembles its surroundings. Whatever the colour of the background is, that will be the colour of the shrimp during daylight. But as night comes on these cells arrange themselves in such a fashion that the shrimp becomes a light, somewhat Cambridge blue all over, no matter what is the colour of its background. These pigment-cells exhibit a daily periodicity corresponding with light and darkness.

Similarly the organisms which cause the phosphorescence of the sea react only when it is dark. During the day they show no light, but if disturbed during the night they emit flashes of phosphorescence which can be readily seen from the bows of a steamer or when the oars of a boat beat the water. Samuel Pepys mentions this phenomenon in his *Diary*. He noticed "the strange nature of the sea-water in a dark night, that it seemed like a fire upon every stroke of the oar." One of the chief causes of the luminosity of the sea is a unicellular protozoon known as *Noctiluca*. Like the pillar of cloud which led the Israelites, it becomes bright by night and dull by day. If you capture some of these little creatures and keep them in an aquarium in a photographer's dark room you would see that they became phosphorescent after night-fall, even though there were no change in the illumination of the dark chamber. These animals seem to remember the time to glow and the time to remain dull in their original more natural surroundings.

Rhythm is often very well marked in cell-division. In the developing eggs of certain animals the various cells divide at almost exactly the same instant for the first six or seven divisions. This results in the formation of a perfectly symmetrical hollow ball of cells.

RHYTHM

If we look at the internal structure of cells during division we find that certain structures (such as the *Aster* in the cell) recur periodically. Dividing cells have recently been dissected under the microscope with the aid of very fine glass needles. They show that a rhythmic change occurs in the degree of the solidity of protoplasm during cell division.

RHYTHM IN TISSUES

Many tissues also exhibit rhythm. The cylinder of growing cells which lies under the bark of a tree increases and multiplies during the summer and is inactive during the winter, thus causing the annual rings which are so conspicuous in a cross-section of a trunk. This is a yearly rhythm. This seasonal growth depends on the alternation of summer and winter. When the leaves fall the trees cease growing. But in the tropics, where the conditions are practically uniform throughout the year, the trees make no such annual rings. Yet if you transplanted deciduous trees, that is, trees that lose their leaves at the onset of winter, to the tropics, they would still continue to show the annual growth, at any rate for several years, and it takes some time before they lose the habit of dropping their leaves in the autumn months.

FIG. 46. Transverse section of an Oak-trunk, 25 years old, showing as many annual rings. From Le Maout and Decaisne.

The curious plant known as the telegraph-plant, mentioned in the preceding Chapter, has slender leaves which rotate periodically at short intervals of time. This rotation is brought about by certain cells accumulating water and swelling up, thus producing pressure in a certain direction. This process passes on to neighbouring cells at the base of the leaf-stalk, and thus causes the whole to rotate slowly. Another example of tissues in plants which show a rhythm is the growing point of a young shoot. If it be carefully observed it will be found that the shoot does not grow straight upwards, but that as it gets longer and longer it

RHYTHM IN TISSUES

makes a spiral movement through the air. This circular movement which becomes a spiral is caused by a rhythmic change in the growing cells. Certain of them will grow faster than the others and then assume their normal rate. Behind them another set of cells will now start growing faster, and this is continued until the circle has been completed.

The growth of plants is not continuous, but is perpetually interrupted in a rhythmical manner. The growth of the *Spirogyra* proceeds by fits and starts; there is a period of activity followed by a period of rest, the intervals extending over some minutes, usually about twenty. Very refined measurements have shown that plant growth takes place by minute but perfectly rhythmical pulsations at intervals of a few seconds of time. The stalk of the crocus, for instance, grows by little jerks, each with an amplitude of 0·002 mm. at periods of twenty seconds or so; and after each little increment there is a partial recoil.

In many forms of movement there is order in time as well as in direction. We have seen this in the case of cilia, and we find it again when waves travel along muscular fibres or from one fibre to another. The muscular movement of the stomach and the alimentary canal which forces the food along is termed *peristalsis*. A constriction takes place in one part of the tubular intestine and this constriction passes rhythmically downwards, and this it does independently of the nervous system. This contraction of the walls of the alimentary canal which is propagated from muscle fibre to muscle fibre takes place at the rate of about 5 cm. per second.

A further rhythm connected with digestion occurs in the cells of the pancreas, the liver and the intestinal glands. In a fasting dog they secrete their digestive juices for a period of twenty to thirty minutes each hour, and then have a rest. Another instance of involuntary muscles which possess the property of rhythmic action is shown in the lower end of the frog's or tortoise's heart. These can be removed from the body and kept pulsating under certain conditions for many days, removed from all nervous stimulus. The heart of the embryo chick begins to beat before any nerves have grown into it. The theory that muscles can contract without being

RHYTHM

stimulated by nerves is known as the *myogenic* theory. The blood capillaries show a rhythmic constriction and dilatation, and still another example of rhythmic contraction of vessels is found in the wings of bats. Here the blood vessels have minute valves which permit the blood to flow only in one direction, and it is driven along by the vessels pulsating at regular intervals.

The rings which one finds in so many natural objects indicate periodicity of growth. We have referred to the rings shown by a section of a tree. The concentric spheres which build up the starch-grains which form so permanent a feature in plant life are believed to indicate a diurnal and nocturnal periodicity of activity and rest, although this may not be the full explanation. Both the scales of fishes, their vertebrae, and the curious bone in the ear termed the *otolith* have marked concentric rings. The subject is a debated one, but many believe, and indeed it seems fairly certain in the scales of many fishes, that these rings indicate an annual growth comparable with the annual rings of the higher plants.

Fig. 47. Otoliths of Plaice, showing four zones or "age-rings." After Wallace.

If you cut a pearl in half and look at the flat surface you will see a series of concentric rings. The cut passes through a series of concentric spheres. This concentricity indicates a periodicity of growth due to the deposit of a mineral, *aragonite*, of which pearls are formed, taking place at intervals. The shells of molluscs also show concentric lines of growth.

RHYTHM IN ORGANS

Many organs of the body act rhythmically. A notable example of this is the heart in man. The duty of the heart is to pump the blood round the body carrying with it oxygen and food to be distributed to the general cells of the body, and bringing back on its return carbon dioxide to be discharged through the lungs, and certain excreta to be discharged by the kidneys and sweat-glands. In a normal

PEARLS AND SHELLS

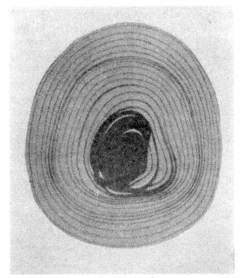

FIG. 48. Section of a Pearl showing concentric layers secreted round a nucleus. From Rubbel.

FIG. 49. Shells of a Fresh-water Mussel, *Anodonta mutabilis*, showing concentric lines of growth.

140 RHYTHM

healthy adult human being the heart contracts about 72 times a minute, but its frequency is interfered with by many factors—exercise, time of day, atmospheric pressure, temperature, food and drink. There is a steady diminution of the number of contractions as age advances. Just after birth the contractions recur at the rate of 140–130 a minute. These drop in the second year to 115–110 and in the seventh year to 90–85. The contractions of a boy of 14 years will be 85–80 and in adult age 80–70. This, however, drops to 70–60 in old age.

Other very important organs which contract and expand rhythmically are the walls of the chest. The number of respirations at rest in a normal human being is about 14 to 18 per minute, 1 to 4 or 5 in proportion to the heart beats.

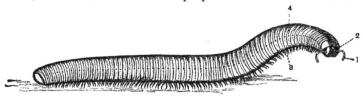

Fig. 50. *Iulus terrestris*, sometimes called the "Wire-worm." From Koch. Magnified about 3½. 1. Antennae. 2. Eyes. 3. Legs. 4. Pores for the escape of the excretion of the stink-glands.

If the heart beat be quickened, the rhythm of the lungs is accelerated. Like the heart, the frequency of the respiratory movements is greatest in infancy and childhood.

If you turn up a stone lying in a damp wood or dig a little in a rich soil you will very probably come across a little black millipede, known as *Iulus*. It is unfortunate that this is known to the farmers as a wire-worm, because they call the larva of a beetle (*Elater lineatus*) by the same name. *Iulus* does a considerable amount of damage by nibbling the tender roots of growing plants. Like other millipedes, it has an extraordinary number of legs. In the group to which millipedes belong, MYRIAPODA, the number of legs sometimes amounts to 150—200. When the *Iulus* is crawling along you can see a succession of waves passing along the legs on each side of the body in a definite rhythm, just as one sees waves pass over a cornfield when the wind is blowing. It is not easy to

LUNAR RHYTHM

determine the sequence of the movement of the legs and, as the following verse shows, it may be a matter of considerable doubt to the millipede:

> A centipede was happy quite
> Until a toad in fun
> Said, "Pray which leg moves after which?"
> This raised her doubts to such a pitch,
> She fell exhausted in a ditch,
> Not knowing how to run.

A very peculiar rhythm has recently been investigated in a sea-urchin in the Red Sea. It has long been a tradition that the moon exercises a peculiar effect on both plants and animals; and apparently this has proved to be true as regards certain marine invertebrates. Aristotle tells us that the ovaries of the sea-urchin acquire a greater size than usual at the time of full moon. Cicero notes that oysters and other shell-fish increase and decrease with the moon. Pliny makes a similar observation, and the same belief is common to-day in many of the Mediterranean fish-markets. A Neapolitan fisherman will:

> Pitch down his basket before us,
>   All trembling alive
> With pink and grey jellies, your sea-fruit;
>   You touch the strange lumps,
> And mouths gape there, eyes open, all manner
>   Of horns and of humps.
>
> BROWNING, *The Englishman in Italy.*

Edible molluscs, crabs, and sea-urchins—"frutti di mare"—are stated to vary with the phases of the moon. When the moon is full the animal is said to be "full." When the moon is new the animal is "empty." Quite recent observation has shown that at any rate in the sea-urchin known as *Diadema setosa*, a very large form, a foot or more in diameter, found at Suez, the testes and the ovaries, which are the edible parts of sea-urchins, show a rhythmic growth and decline corresponding with the lunar cycle. At full moon these generative organs are at their largest, and then spawning takes place. After this the testes and ovaries grow smaller, and after the new moon they begin to increase again in size in preparation for a further season of spawning. The sea-urchins of the Mediterranean do not show a similar lunar rhythm and the popular belief that they do may possibly be due to the fact that in very early times the Egyptians, exceptionally able people, noted the fact that the Red Sea urchins had a lunar period, and generalizing too widely spread the view amongst the Greeks and Romans.

The influence of the moon on the food supply of the people is shown by the "Grunion," a small smelt some 6 inches long which occurs on the sandy shores of California on the second, third, and fourth nights after the full moon, during the big tides from March till June. These little fish come ashore on the top of a wave, lie for a moment on the sand, and drop back into the sea with the succeeding wave. Hundreds of people assemble to pick up and catch these fish in the moonlight, using screens, seines, and their hands. The object of the people is of course to get food, but the object of the fish is to spawn. The eggs are laid in small masses $3\frac{1}{2}$ inches below the surface just above the limit of the average tide. They even form the food of a small terrestrial beetle. Ten days after spawning and one day before the next high tide the eggs are washed out of the sand, hatched, and the larvae escape into the surf. The whole procedure is wonderfully timed and adjusted so that the fish have just enough time on land to lay and bury their eggs. The "run" of the fish begins about the turn of the tide. Otherwise, eggs laid an hour too early would be washed away.

RHYTHM IN ORGANISMS

Bacteria and Protozoa exist in the soil, as we have seen, in incredible numbers. They do not remain the same in number from day to day and the fluctuations are great, but, so far, have not been associated with any definite external factors. They do, however, show a certain fortnightly rhythm, and both bacteria and protozoa which prey upon bacteria are most numerous at the end of November and at a minimum during February. But so far these changes have not been associated with temperature or rainfall.

Amongst the numerous organisms which were investigated at the Rothamsted Agricultural Experimental Station every day for a whole year is a flagellate, known as *Oikomonas*, and this shows a curious two-days rhythm which is at present quite unaccountable. Every other day it swells up and appears in greater numbers and on intermediate days it diminishes. Why it does this is quite unknown.

Another daily periodicity is shown by man both in his weight and height. The weight is lowest in the morning before breakfast, and highest after the evening meal. This is easily explicable. But the variation in height is rather peculiar. During the day the stature decreases from 1 to 3 centimetres. This is attributed to the compression of the intervertebral discs whilst man is in an upright position, to the curving of the spine, and to the depression in the arch of the foot. It is therefore necessary in taking measurements for comparative purposes to take them at the same time of the day, preferably before breakfast.

Many aquatic organisms show a periodical ascent and descent in the water in which they live. This is attributed by some to variations in the temperature, in the strength of the light, and in the food supply. One of the chief factors is that many floating animals move towards the bottom when light falls on the water, and upwards when darkness sets in. But another factor seems to enter into this periodical rise and fall. It has been known for some time that the little water-flea *Daphnia* increases in specific gravity after a meal. There is a marked difference in its specific gravity twelve hours after feeding and after twelve hours' starving. Recent observations have shown that its maximum density is normally acquired between the hours of 6 and 8 a.m. By noon there is a less well-defined but still marked minimum specific gravity. *Daphnia* takes an early and hearty breakfast. The minute algae upon which it lives are most abundant near the surface in the early morning. When the *Daphnia* rises from the deeper water it finds itself surrounded by an abundant food supply. After a meal it seems to sink slowly. There is a periodical rise and fall of the animal towards and from the surface layers. *Daphnia* is most scarce in shallow water from midday till about 4 a.m., and direct sunlight also seems to drive it downwards. There are, of course, other factors. The number of young in the brood-pouch is one of these. When the brood-pouch empties itself the specific gravity temporarily, at any rate, decreases.

Here we have a diurnal rhythm probably depending upon the variation of the food supply, and this is probably dependent upon such factors as light and temperature.

The little fresh-water Rhizopod *Arcella* also shows a certain rhythm. It sinks gradually, and then, feeling the want of oxygen in the depths of the water, secretes from its own protoplasm a bubble of oxygen which, acting like a float, brings the protozoon to the surface again.

Light and darkness, the alternation of the seasons, produce a rhythm which affects nearly all plants and animals. The evening primrose opens as twilight comes on; the mosquito becomes active at the same time, retiring into its base before the sun rises.

Most of the animals on the sea-shore are subject to the rhythm of the tides. There is a little green worm, called *Convoluta*, just visible to the naked eye, which occurs in such numbers that when they are above ground great splashes of green are observed on the sandy beaches of the coast of Normandy and Brittany, near the level reached by high water. As the tide laps up, these animals disappear into the sand. Twice every twenty-four hours the colonies are submerged and live in darkness underground, and twice every twenty-four hours the sand is uncovered by the sea and the green animals emerge. Their burrowing away seems to be due to their desire to escape the wave-shock, and their returning is due to the fact that they contain many green algae which require the light of the sun for their assimilation. The egg-laying is similarly influenced by the spring tides, the condition most favourable to egg-laying occurring when the moon is full or new. A curious feature of these animals is that if taken away from the sea and kept in an aquarium inland they still, at any rate for a time, sink when the tide is rising at their far-away home, and emerge when the tide is flowing out. This habit is ingrained in them.

Periwinkles show also the influence of the tides. When the tide is down they are inert and inactive, but can be made to renew their activity by shaking. If we take these periwinkles

DIURNAL AND LUNAR RHYTHMS

inland they are more difficult to shake out of their inactivity at periods when the tide is low than at periods when the tide which they have now forsaken is high. Their period of rest and sluggishness in the laboratory corresponds to the period of drying up on the seashore.

A curious example of night and day influence on the periodic movement of birds and mammals is recorded on the coast of Ceylon. Off this coast is a small island frequented by night by crows, and by day by fruit-eating bats or "flying-foxes." The crows fly to the island to rest in its trees at night. In the morning they return to the mainland to feed. The bats fly to the island as daylight dawns to rest for the day suspended upside down in rows from the palm leaves. Thus, like Box and Cox in the well-known farce, the same bedroom accommodates both day and night sleepers.

FIG. 51. *Nereis pelagica*, a worm belonging to the same sub-order as *Eunice* and *Odontosyllis*. After Oersted.

Perhaps one of the most astonishing examples of the state of the moon influencing the movements of animals is afforded by the marine worm known as the Palolo-worm, *Eunice*, which lives in the coral reefs of many a Pacific Island. At the last quarter of the moon in October and November the hinder part of the worm, which is crowded with reproductive cells, breaks off from the foremost part, the latter remaining amongst the corals, and the hinder part floating to the surface of the sea, where it gives exit to its eggs and spermatozoa. After shedding the reproductive cells this half of the worm dies. The spawning of this half of the worm takes place at low tide on several successive days; and it is eagerly sought after by the native fishermen, who esteem it a great delicacy. So regular is the appearance of these worms that the fishermen know the right time to prepare their boats and nets, and so far they have never been disappointed. There are several other worms of a somewhat similar nature, such as certain marine worms in Japan and in the Atlantic, which show a similar rhythm.

A somewhat similar state of things has been observed in a marine worm known as *Odontosyllis enopla* which occurs in the Bermudas. It reproduces three times a year, at the first low tide in July, the moon being in its last quarter; then again 26 days later and in less abundance; and after an interval of 26 or 27 days, *i.e.* late in August. They appear on the surface at dusk, usually within five minutes of eight, and the phosphorescent display is always over by half-past eight. At first the female, swimming on the surface of the sea, shows no light, but quite suddenly she becomes markedly phosphorescent, swimming through the water in small luminous circles. Around this circle is a halo of phosphorescence. If the male does not appear the illumination ceases after ten or twenty seconds, though it may be repeated four or five times. Usually, however, the males are abundant, and unmated females are rare. The male appears as a glint of light at first some ten or fifteen feet from the luminous female. It comes up obliquely from the deeper waters and darts at the centre of the luminous circles, finding the female with remarkable precision. The two then rotate, performing a kind of ceremonial dance, scattering and shedding eggs and sperm in the water.

Then again, that curious mollusc with its segmented shell, the *Chiton*, breeds only at the time of full moon in certain parts of the world. It has even been said that the number of births in large human communities show a certain correlation with the sidereal lunar period of 27·3 days. This is the time taken by the moon in making one circuit of the heavens.

It has been pointed out that during the last century and a half the maximum and minimum herring fisheries have coincided with the maximum and minimum declination of the moon, which occurs at intervals of just over nine years. Both in Denmark, in Italy and in Egypt the eel fishermen maintain that their best catches are always taken when there is no moon; and it is on record that strong lights will check or retard the migration of these fishes up the rivers.

PLANT RHYTHM

The effect of the moon on plants has been chronicled much more rarely, but there is a common alga known as *Dictyota* which in one part of the world produces its sexual cells once in each lunar month, and in other parts of the world twice a month.

Certain flowering plants are said to bend their stem towards the moon and follow its course by night as it moves on its fixed path. The effect of the seasons on many perennial plants is too well known to need much comment. They burst into life as spring comes on, live throughout the summer, and their visible parts disappear with the autumn, though tucked away underground is a remnant capable of reproducing the plant at the return of spring.

> This season's Daffodil,
>   She never hears,
> What change, what chance, what chill,
>   Cut down last year's:
> But with bold countenance,
>   And knowledge small,
> Esteems her seven days' continuance
>   To be perpetual.
>   KIPLING, *Cities and Thrones and Powers.*

The climatic changes in the tropics are at a minimum both in the hot, damp forests and on the desert tracts of Africa, Asia and America. The changes of season are but very slightly marked. Hence the activities of plants and animals proceed all the year round without the interruptions caused by the hot and cold seasons of more temperate zones. Trees in the tropics make no annual growth-rings. Nevertheless there is a periodicity, a rhythmic alternation of periods of repose and activity. Certain plants and animals as a whole show little of this periodicity, but even these have resting periods for certain of their functions. The less the periodicity of the climate is, the less is the plant dependent on it, and such alternations of rest and activity in a uniform climate are in the main due to internal causes. In the densest and most humid tropical forests there is a time for breeding and a time for vegetative recovery. But this may not affect all the plants of one species at the same time, and hence one finds that a given species is in bloom for long periods of time even during the whole year, although each individual specimen may only flower for a few days, or, at the outside, for a few weeks. Where the climatic change is slight, trees often shed their leaves at longer or shorter intervals, sometimes as often as six times a year, sometimes only once; and this process is independent of the seasons of the year, so that trees of the same species under the same conditions drop their leaves and acquire new ones at times that do not coincide. This is carried in some cases even further, for we find that in certain trees, for example the orange-tree, the individual branches have become independent of one another, so that on the same tree winter, spring, summer and autumn shoots may be found on different branches. Flowers and fruit may be found at the same time on a tree but on different branches. Plants that live in desert areas have a special adaptation for storing water, and they show a certain irregular periodicity in their size, which is dependent on rare and infrequent rainfall.

Although there is no winter or summer, spring or autumn, in the tropics, both plants and animals are subject to the periodical rise and fall of the sun and the monthly waxing and waning of the moon. But there is another immense area—and a very densely populated area—of the habitable globe, where even these periodicities are eliminated. In the depths of the sea, two or three thousand fathoms below the surface, we find a very large population of marine animals, and the conditions under which they live are extraordinarily uniform.

Below about three hundred fathoms the light and heat of the sun hardly penetrate. Hence, no green plants can live below this limit. Diatoms and algae, which form so large a proportion of the living matter at the surface, cannot live in the absence of sunlight. But the depths are peopled by very large numbers of species of animals of all sorts, and no part of the sea contains a denser population. These deep-sea animals live to some extent on each other, but like other creatures they cannot be self-supporting. They cannot subsist as the inhabitants of the Bermudas, who were said to subsist by "taking in one another's washing." Like the inhabitants of great cities, the dwellers in the depths must have an outside food supply, and this ultimately comes from the surface or

RHYTHMS IN THE DEEP SEA

from the numerous animals which, living in the middle waters, die and fall to the bottom. Others, members of the middle region fauna, again, may swim down there and be caught. There is a wonderful uniformity in the state of things at the bottom of the deep blue sea. Climate plays no part in the life of the depths; storms do not ruffle their inhabitants; these recognize no alternation of day or night; seasons are unknown to them; they experience no change of temperature. Although the abysmal depths of the polar regions might be expected to be far colder than those of the tropics, the difference only amounts to a degree or so—a difference which would not be perceptible to us without instruments of precision.

At the bottom of the sea there is no sound—

There is no sound, no echo of sound, in the deserts of the deep,
On the great grey level plains of ooze where the shell-burred cables creep.

The world down there is cold and still and noiseless. The numerous inhabitants of the depths are uninfluenced by the daily rise and setting of the sun, the monthly movement of the moon, the succession of the seasons. Each of them might exclaim with our great epic poet

> Thus with the Year
> Seasons return, but not to me returns
> Day, or the sweet approach of Ev'n or Morn,
> Or sight of vernal bloom, or Summer's Rose.

Nevertheless they also show a certain rhythm, and although little is known about their habits it seems certain that they spend part of their time building up their reproductive organs, their eggs and their spermatozoa, and this is followed by the shedding of the same, to be followed in its turn by another period of what is termed vegetative growth. But what rhythm they do show must be an internal one and not dependent on outside factors.

Although there appears to be a total absence of external rhythms in the depths of the sea, such rhythms are extremely pronounced at the surface, and, indeed, for some little distance beneath the surface. There is the rhythm of the tide, a rhythm which corresponds with its rise and fall about twice every twenty-four hours, and with this is involved a still bigger fortnightly rhythm corresponding with the full and new moon; for about half-way between these two phases the tide rises more slowly and to a lower height. And again, just as there is a half-daily and a half-monthly rhythm, so we have a half-yearly rhythm in the vernal and autumn equinoxes. So regular are these rhythms that the tide is calculated years in advance for all parts of the world, and navigators rely trustfully on these calculations, which are not found wanting.

Then, again, we have a rhythmical change of temperature, which is fairly constant for given places in the sea. About February and March the sea in the northern hemisphere is at its coldest, but it gradually warms up until in August it attains its highest normal temperature. Of course, in all these rhythms there are many disturbing features, such as the weather. But these can fairly easily be discounted. Just as we have the annual rise and fall of temperature, so do we have a daily one, the temperature being at its lowest about sunrise and gradually rising till about the middle of the afternoon. And again, there is a fortnightly rhythm, inasmuch as near the land the sea is warmer in the summer just after the time of new or full moon, and colder at the same periods during the winter.

Other rhythms might be pointed out, such as those dependent on the intensity of sun-light, and on the degree of salinity, which in turn depends to a very large extent on the water circulation of the sea. The pulsing-up of the Gulf Stream is the direct result of this circulation and affects not only the warmth but the salinity of the waters on our western shores. "The water is saltest when the drift is strongest, in the months of February to June, and is less salt when the drift is weakest, in the months of November to February." All these features have a profound influence on the life of both plants and animals in the ocean.

There is no part of the sea which is free from animal life; but it wells up and dies down periodically at intervals during the twelve months. On the surface of the sea there are numerous animals which are found there and only there, such as the jellyfish, numerous small crustacea, floating molluscs, worms and others. But many marine animals pass only part

RHYTHM OF THE SEA

of their life near the surface, and as a rule it is the larval stages which exist in countless quantities at and just below the sea-level. We can trace a distinct rhythm in this part of the ocean. It has a regular succession of organisms during the twelve months, and the skilled zoologist could make a very shrewd guess at the time of year at which a collection of floating organisms was captured. At the end of winter and the beginning of spring in our seas there is a great outbreak of diatoms, those unicellular algae which form the ultimate food of fishes. As the sunlight increases and the amount of food in solution becomes larger, diatoms multiply and increase greatly. A little later the fish begin to spawn, and copepods and other small crustacea increase greatly in numbers. Soon the larvae of many deep-sea and middle-sea forms begin to appear in the tow-net. The larvae of barnacles come to the surface in enormous numbers as spring sets in. These may linger as late as April or May and some even until the end of summer. In March we find the crab spawning, and its larvae soon become prominent. By May the number of diatoms begins to decrease, and this decrease lasts for some few months. Soon the summer-spawning fish produce their surface larvae, and so do the sea-urchins and starfish. The phosphorescence is perhaps at its height in July, and during the summer months swarms of jellyfish stain square miles of the sea a uniform reddish colour. In the late autumn and winter months the diatoms begin to reappear and reach the second peak of their curve in November, after which month they die down again. As winter advances, the surface fauna and flora fall away both in quantity and in variety; still some animals and plants are persistent the whole year round, and take no part in this cycle.

There is thus a regular succession of organisms in the surface waters, and it has been pointed out that the main features of this succession are: (1) the relative scarcity at the beginning of the year; (2) an outburst of diatom life in the late winter and early spring; (3) the appearance of countless myriads of fish and invertebrate eggs and larvae in the spring and early summer; (4) a decrease in the abundance of the diatoms, and the gradual disappearance from the *plankton*—the drifting,

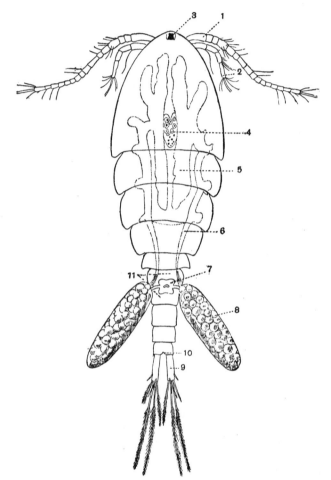

FIG. 52. View of female *Cyclops* sp., a typical Copepod. Magnified. After Hartog. 1. 1st Antenna. 2. 2nd Antenna. 3. Eye. 4. Ovary. 5. Uterus, *i.e.* pouch of the oviduct into which the eggs pass before being shed. 6. Oviduct. 7. Spermatheca or pouch for receiving the spermatozoa of the male. 8. Egg-sacs. 9. Caudal fork. 10. Position of anus. 11. A segment, bearing the genital opening.

RHYTHM OF THE SOIL

and for the most part, surface organisms—of the eggs and larvae; (5) the appearance of swarms of medusae and other coelenterates in the summer; (6) the reappearance of diatoms in the late summer and autumn; and (7) the scarcity of the plankton as the winter begins. Throughout all these changes a number of forms of life remain as permanent inhabitants of the surface area all the year round.

The temperature of the soil also undergoes a daily and yearly rhythmic change, at any rate within certain limits of depth from the surface. At the surface there is a daily rise and fall of temperature, and this temperature wave is propagated into the soil and as a result the temperature at any given depth shows a fluctuation which is a kind of reduced image of that of the soil surface. The time between the maxima and minima of temperature remains unaltered. But the temperature range becomes diminished, and there is a "time-lag" which at a depth of six inches may amount to some hours. A rise of temperature which begins at the surface soon after daybreak is not apparent at a depth of six inches until about 9 a.m., after which it becomes warmer and warmer till about 3.30 p.m., and then remains constant for an hour or thereabout. About 5.30 p.m. it begins to fall slowly and continuously throughout the next sixteen hours, that is to say, till about 9 a.m. the following day.

Soil warms up more quickly than it cools. It attains its maximum temperature in seven hours, and it remains constant for about an hour and takes about sixteen hours to cool. These figures, however, vary to some extent at different seasons and the daily rise is slower after rain, though wind appears to influence it but slightly; and there are other factors which slightly interfere with the figures. Roughly speaking, the soil population enjoys a warmer climate than that of the atmosphere, and a moister one. The minimum temperature of the soil is in the summer from six to eight degrees Centigrade above the minimum temperature of the air, and in winter some three degrees higher. Further, the minimum temperature in the soil is attained early in the day during the summer, somewhere about 7.45 a.m., and in the winter when it is coldest about 10.30 a.m. The air on the other hand is coldest at the same time of the day, about 3.45 a.m., all the year round.

There is a curious relationship between the occurrence of infantile diarrhoea and the temperature of the soil. This very fatal disease is conveyed by flies which foul the food of infants; but it has been noticed that the disease does not reach epidemic proportions unless and until the temperature of the earth at a depth of 4 feet is about 56 degrees Fahrenheit. In 1922 the temperature of the soil at this depth only just reached this figure and there was no epidemic; but in the late summer of the following year, 1923, the deeper temperature stood at 58·9 degrees, and infantile mortality due to diarrhoea rose to a figure higher than any recorded since the autumn of 1921.

Rhythm in Communities

Merely because man happens to be a vertebrate, we are rather apt to think that the Vertebrata are the most dominant group of animals in the world. But in certain respects the great group of insects can claim that position, at any rate as regards land surfaces, for they are almost entirely absent from the ocean. In number of species they surpass all other terrestrial animals; compared with the vertebrata their number is colossal. It has been stated that the greater part of the protoplasm of the world is locked up in the bodies of insects. They have further developed a social system, an art of living together in societies, more perfectly organized than those which our civilization has succeeded in producing amongst men; and they have actually achieved a high standard of industrial and agricultural activity which facilitates their social life. Familiar examples of these communities with individuals specialized for the different kinds of work of the colony or society needs are termites or white-ants, wasps and bees, and perhaps the highest of all, ants. The first named, whose white ant-hills sometimes attain a height of twenty feet, are for the most part inhabitants of tropical or sub-tropical countries, and the seasons have no effect upon their activities. In the same way the real ant communities carry on, but little affected by changes of temperature, even in their most northerly

RHYTHM OF BEES

and southerly geographical limits. On the other hand, wasps and bees in temperate climates show a marked periodicity. Whereas in the tropics the bees' nest or the wasps' nest is in action throughout the twelve months, in a temperate climate these activities are reduced to a minimum during the winter. The honey-bee is alone amongst bees in maintaining its colony throughout the winter; but the activity of the hive is greatly reduced. The workers stop at home; they cease to collect honey and pollen, and live on what they have already stored up. The temperature of the hive, always well above that of the surrounding atmosphere in summer, drops, but is still higher than that of the outside air. The drones are murdered and cast out of the skip; but although the activity

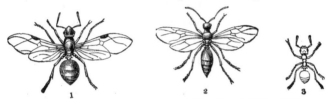

FIG. 53. The Wood-ant, *Formica rufa.* 1. Female. 2. Male. 3. Neuter.

of the workers is reduced, they do not cease to work, for, like Martha, they are "cumbered about with much serving."

This marked seasonal periodicity is even more pronounced in the case of the bumble-bee and the wasp, for with these the whole colony dies out at the commencement of the winter with the exception of the queen. A bumble-bee's nest is in being only for about three or at the most four months of the year; for the remaining nine or eight months the whole potentiality of the next summer's brood is hidden away in the body of a fertilized queen.

If we try to trace the history of the bumble-bee's nest we may begin with the queen in the late summer. The active season of such a nest is shorter than that of the *Apis* or *Vespa*, the closed time longer. The final activity of the corporate life is the rearing of queens in the later part of July or in August. Once grown up, the queen as a rule soon leaves the nest, but she is "a shy bird" and hides herself away in some cranny or among some débris. Here she is diligently sought for by the males, who pause at every likely spot and emit a very pleasing scent, possibly to attract the queens. Sometimes they try to intercept their brides as they leave the nest, but in any case the queen, once fertilized, abandons her home, which soon falls into decay, and seeks for winter quarters. Before leaving the nest she has filled up her crop with honey, and this must suffice her for food during the next eight or nine months, when she is *en retraite*. The queens of some species, *Bombus terrestris*, like to winter in burrows under trees, those of others, *B. lapidarius*, prefer crevices high up in banks. But whatever habitat is chosen, damp must be avoided and the aspect must be northerly. This latter is also true for hibernating wasps, and the explanation is not far to seek. These burrowing females are roused to activity by the warm spring sunshine; should their winter home face south, a single exceptionally warm winter's day might awake them. They would emerge to find the world unready for them and perish without founding a colony. Getting up too early with them is as fatal as not getting up at all. The retreat of the queen, at any rate in the species *B. lapidarius*, is often revealed by little heaps of sand or earth, excavated as she tunnels the bank to a depth of two or three inches. At the end of the tunnel she carves out a spherical cell an inch or more in diameter. At first she sleeps but lightly, and if disturbed by any cause will emerge from her "cell" and fly away to build another; but as winter approaches and the days become cold, she sinks into a deep lethargy, simulating death. This torpor lasts eight or nine months. Those species who go to bed early begin to stir as early as March; those that retire later may not resume their activities until May or even June.

As the spring advances the queens re-appear, and "may be seen busily rifling the peach blossom, willow catkins and purple dead-nettle." At first they nightly retire to their hiding-places, but soon, as the days lengthen, the desire for starting the colony becomes irresistible and a home is sought out, usually one already made and abandoned by some field mouse or other small, burrowing mammal.

RHYTHM OF WASPS

As autumn approaches, the wasps' nest, like the nest of the bumble-bees, disintegrates and disappears, while the hive of the honey-bee persists throughout the winter, bereft only of its drones, which are usually done to death by the workers. Two categories, the queen and the workers, live through the winter in the honey-bee colonies, with lowered activities, but yet alive and ready to resume work when spring arrives. The hive also remains, the comb is not destroyed, and after the worker bees have carried through a little spring-cleaning—so dear to the female heart—it will be ready for use again when next needed.

But with the wasps' nest things are far otherwise. As the autumn approaches and the cold weather comes on the young

Worker-bee. Queen-bee. Drone.
FIG. 54. The Honey-bee, *Apis mellifica*.

queens, which have previously been fertilized by the drones, retire from the world and hide away in moss or under a thatched roof or in some corner of a shed. Here they seize with their jaws some fragment of straw or bit of rag and, hanging on almost entirely by their jaws, wrap their wings round them like a dressing-gown and enter on their winter sleep. Meanwhile the wasps' nest has been rapidly deteriorating. The activities of the workers fall off, the drones are slain or cease to return to the nest. The workers sustain life for a few more days by devouring the remaining larvae and pupae, but soon they also perish. The nest begins to crumble, and so the ruins of what was the home of one of the most highly organized of insect communities serve but to shelter field mice, earwigs, mites, beetles and woodlice. Seen under a diminishing glass,

> the Lion and the Lizard keep
> The Courts where Jamshyd gloried and drank deep.

It thus comes about that the whole wasp protoplasm, or the real living matter, is throughout the winter months tucked away in the bodies of the queens which are hanging by their teeth in some remote cranny. In her body, and in her body alone, is the potentiality of the wasps' nest, and should she perish next year's wasps' nest perishes with her. This, indeed, often happens, especially during a mild winter with snaps of really cold weather.

Similar periodicities occur in human societies. Great civilizations have been time after time evolved. From the

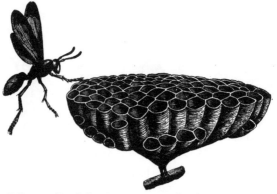

FIG. 55. *Polistes tepidus*, belonging to the same family as the Wasp, and nest.

humblest origins they have reached a height which we to-day are only just beginning to appreciate; then gradually they have crumbled away and disappeared. The civilizations of Assyria, Babylon, Egypt, Crete and Arabia, the great societies of Mexico and Central America have passed away, and the story of their passing serves to-day as a reminder of the fate which has hitherto overtaken all human civilizations. One of these days they may, in the cycle of time, recover and enter upon a new period of prosperity, as Italy has done after the collapse of the Roman Empire. But the laws which govern the periodicity of civilization have yet to be discovered. The great French and English historian Hippolyte Taine believed that the laws of civilization may be discovered by a scientific analysis of History, but may it not be possible that when discovered they will be found to be in the same category as those laws of rhythm of whose operation in Nature we are now slowly becoming aware?

In concluding this long chapter it is just worth while to point out that the highest and most artistic pleasures in human life are due to causes which are essentially rhythmic, light and colour, music and poetry. The rhythm of the human voice, of a trained orchestra, the rustling wind in the trees, the changes of light during a summer's or even a winter's day, the colour of the skies, appeal in a greater or less degree to all mankind, and afford an amount of pleasure which is beyond all expression or even calculation.

CHAPTER XIII

REPRODUCTION

GROWTH AND REPRODUCTION—SPORES—VEGETATIVE REPRODUCTION—SEXUAL REPRODUCTION—ALTERNATION OF GENERATIONS—NUMBER OF OFFSPRING—ORIGIN OF LIFE FROM THE SEA—THE EGG—ANTHEROZOIDS AND SPERMATOZOA — HERMAPHRODITISM — PARTHENOGENESIS — OLD AGE AND DEATH

> Because, in point of fact, one does see, in this world—which is remarkable for devilish strange arrangements, and for being decidedly the most unintelligible thing within a man's experience—very odd conjunctions.
>
> Cousin Feenix. *Dombey and Son.* CHARLES DICKENS.

GROWTH AND REPRODUCTION

ONE of the distinctions between living organisms and inorganic matter is that the former grow and reproduce and the latter does not.

It is quite true that crystals and other inorganic substances get larger. But in their case the growth is due to new matter being deposited outside them. Like Mr Weller, senior, when he put on one overcoat after another, they increase in bulk by a series of external accretions. But living organisms take up food and, after digesting it, pass it all through their substance, and it accumulates throughout the body, adding to the existing protoplasm or cell-tissues. This manner of growth is termed *intussusception*.

There is, of course, a limit to growth. Trees and elephants are very much bigger than mushrooms and mice. In animals, as a general rule, reproduction occurs at the time that the limit of growth has been reached, though there are many exceptions to this statement; but many plants grow throughout life and reproduce most of the time. Unicellular plants and animals are often more or less spherical. It is well known that the surface of a sphere increases as the square of the radius, whilst its volume increases as the cube of the radius.

SPORES

Thus, organisms as they get larger and larger have a smaller and smaller surface in proportion to their contents, and at length a size is reached when the surface is too small to take in enough oxygen or food for the bulky interior. When this size is reached, reproduction generally takes place.

In simple unicellular organisms the whole body is reproductive. In an *Amoeba*, for instance, which is reproducing, the nucleus first divides into two halves. The halves then separate and move apart, and between the two a constriction occurs in the protoplasmic body which, getting deeper and deeper, ultimately separates the *Amoeba* into two *Amoebae*.

> When we were a soft amoeba, in ages past and gone,
> Ere you were Queen of Sheba, or I King Solomon,
> Alone and undivided, we lived a life of sloth,
> Whatever you did, I did; one dinner served for both.
> Anon came separation, by fission and divorce,
> A lonely pseudopodium I wandered on my course.

Spores

The *Amoeba* divided into two equal parts, and neither can be regarded as the parent of the other. The yeast cell also divides, but unequally. A little bulge appears on its spherical surface, which grows bigger and then breaks off. In this case the larger cell may be regarded as the parent. If yeast cells are starved, the protoplasmic contents of each cell divide into four, and each of these four new cells secretes round it a thickened cell-wall. Should the yeast dry up, these

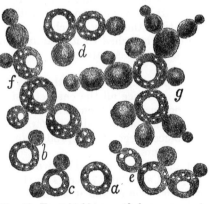

FIG. 56. Yeast highly magnified. *a–g*, successive stages of budding. (Darwin's *Elements of Botany*.)

spherical *spores* escape and are blown about, and should they reach a suitable medium they will soon turn into ordinary yeast cells. These spores, in fact, enable the plant to "carry on" when the ordinary yeast cell would die for want of moisture and food. Many other fungi reproduce by means of spores. The filament of a mould, for instance, may undergo a series of constrictions and come to look like a string of beads. The beads break off one by one and each forms a spore. In other species a number of spores will be formed inside the cellulose cell-wall of the filament. These are termed *endospores*, and when released they have similar powers of resisting adverse circumstances, and they help to spread the plant by being blown about.

Many algae and some fungi that live in damp places produce spores which swim actively through the water by the aid of one or two flagella. These are motile *zoospores*.

Vegetative Reproduction

There is, however, another method of reproduction which occurs in most plants and in many of the lower animals. Parts of the body may separate off from the main organism and grow into complete individuals.

When a plant reproduces itself by simply separating off a part of its body, the process is called *vegetative reproduction*. For a time the body which is going to form the new individual is attached to its parent and nourished by it; but after a time it separates off and becomes an independent organism.

Now plants produced in this way are part of their parent. These reproduce all the characteristics of the parent. They show no variations such as are introduced when fertilization, *i.e.* the union of two cells, *gametes*, each from a different plant, takes place. Vegetative reproduction is very common amongst higher plants.

A Canadian water-weed known as *Elodea* was introduced into the river Cam about 75 years ago by a Professor of Botany at Cambridge whose name it is kindlier to conceal, but the plant was locally known as *Babingtonia abominabilis*. It grew to such an extent that it spread through all our streams and canals and at times choked them. As only the female plant was introduced and not the male, the whole of this extraordinary

VEGETATIVE REPRODUCTION 163

growth is due to the vegetative formation of buds which lengthen into branches and the branches then separate from the stem. Sometimes, as in the case of the strawberry, or some buttercups, *runners* are formed and from the tip a new young strawberry plant grows out. The runner then becomes severed and we have two plants where before we had but one. The *eyes* of the potato are underground buds, and similar buds exist on the Jerusalem artichoke. Here when the buds sprout and grow into new plants the parent body dies; but

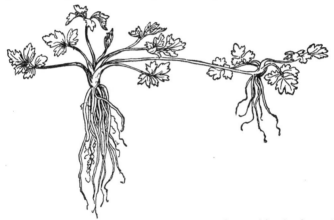

FIG. 57. Creeping Buttercup, showing creeping shoot arising in the angle between a leaf and the main plant. After Praeger.

in other cases, such as the strawberry, the parent survives. In all these plants overcrowding occurs sooner or later. We have also what are called *adventitious buds*. These occur in all sorts of odd places. If a leaf of a *Bryophyllum*, one of the stonecrop or houseleek family, be pegged down in moist, damp earth, little buds will turn up in the notches of the leaf. The leaf then decays and the isolated buds carry on the race. Similar buds are produced in this way by the *Gloxinia*, the *Begonia* and other plants. Roses are capable of forming buds even on their roots, and the fact that the same is true of the dandelion explains the difficulty in clearing a lawn of these pests.

Gardeners use vegetative reproduction by making cuttings or by *layering*, which means pegging down below the soil portions of shoots which have been partially cut through. Carnations, mulberries and many other plants are often reproduced in this fashion.

Vegetative reproduction is particularly common amongst Alpine plants, probably because the short season and low temperatures are unfavourable to the formation of flowers and fruits. Vegetative reproduction is not common among the Gymnosperms (firs, larches, pines), but in ferns, club-mosses, horsetails, etc. it is very prevalent. It is so common amongst mosses that some kinds have never been seen in fruit. It is equally common amongst the Algae and the Fungi. In certain cultivated plants it seems to be the sole method of reproduction. The pine-apple and the banana are often seedless. Sugar-canes rarely flower and the Jerusalem artichoke has been grown in our gardens since the time of Queen Anne solely from tubers. The fundamental difference between vegetative and sexual reproduction is that organisms which are propagated in the former way keep true to type. When you have a good strain you can maintain it; but it does not make for variety.

Vegetative reproduction is much less common amongst animals than plants. In the latter, as we have seen, it is fully maintained even in the highest plants; but in animals it does not get very high up in the scale. Many animals reproduce by dividing, for instance *hydroids* such as *Hydra*. Should the products of the division separate completely, we have two hydroids instead of one; but in many cases the results of the division remain in continuity and then we have a colony, and each constituent forms what is called a *zooid*. Even animals as high in the scale as sea-anemones will divide into two, and the same is true of many worms. If an earthworm be accidentally divided it will regenerate the cut surfaces and form two earthworms; but many aquatic worms divide normally. In the case of the Palolo-worm mentioned on page 145 the hinder end is packed with reproductive cells. This separates off and floats to the surface. The front half, which remains in the depths of the coral rock, then proceeds to regenerate the lost portion. Some worms also throw out buds or lateral branches which remain in contact, and thus, as in *Syllis ramosa*, a marine worm which lives inside a sponge, a regular branching net-work is formed.

The so-called *gemmules* of sponges are formed in the late autumn in temperate climates, and in tropical climates at the beginning of the dry season. Each gemmule is built up of a number of ordinary cells which secrete around them a capsule, the whole resembling a seed. They are capable of withstanding most adverse conditions and can survive severe frosts and the drying up of the water in which they live. They are more common among fresh-water sponges than among marine forms.

The *statoblasts* of the Polyzoa, fairly highly organized animals living in colonies, resemble in general the gemmules of sponges. They again consist of a mass of cells surrounded by a hard capsule and their object in life is to keep the race going when their parent forms have died down owing to frost or drought. The capsules are very characteristic of the species to which they belong. In temperate climates they are formed in the autumn and germinate in the warm spring days.

Sexual Reproduction

We have seen that the simplest form of reproduction is dividing into two. When the two products are of equal size, as is the case with *Amoeba*, the process is called *fission*, but when they are of unequal size, as in the yeast-cells, the process is called *gemmation* and for some illogical reason the larger product is regarded as the mother. A stimulus to this fission is often given by the fusion of two organisms. These may be of similar size, as in the case of *Paramoecium*. Two *Paramoecia* will lie close to one another and interchange some of their protoplasm. This is known as *conjugation* and is often a prelude to a fission. In other unicellular plants and animals sexual spores may be formed, that is to say, the protoplasm divides into 2, or 4, or 8, or more, separate individuals usually in a capsule, from which they escape and then may fuse two by two. A further stage is when the spores are of unequal size, and in that case the fusion of a smaller one with a bigger one is a forerunner of fertilization of the ovum by the spermatozoon.

As we pass a little higher in the scale, we find simple plants and animals with more than one cell, and yet a little higher we find some of these cells turning into ova or eggs, whilst the others divide up and form a large number of minute swimming cells called antherozoids in plants and spermatozoa in animals. The egg is always a large, massive, inert cell, packed with food material for the maintenance of the resulting offspring, whilst the male cell is active and small and capable of chasing the ovum. In these multicellular organisms (often consisting of but a few cells, 8, 16 or 32) not all the cells will become either ova or break up into spermatozoa. There will be others that have been used to take up food, to provide locomotor organs, etc., etc., so that the organisms now consist of reproductive cells and cells which are not reproductive. The latter form the body or *soma*, as opposed to the reproductive cells or *gametes*.

In the unicellular plant or animal which divides by fission there is no difference between reproductive cells and body cells. Such an animal as *Amoeba*, or such a plant as *Bacterium*, are potentially immortal. They divide and divide and divide; but there is nothing to die, there is no body, there is no corpse. Of course they may be killed; but it is at least possible that the *Amoeba* we see under a microscope has had a continuous life since the creation of the world—that act of "impardonnable imprudence," as Anatole France called it.

The whole subject is admirably put by one of our best writers on Natural History, and I venture to quote some paragraphs from his pages.

It was Weismann, with his characteristic habit of pushing ideas to their logical limits, who startled biologists by the conception of the immortality of the simplest organisms—the unicellular Protozoa and Protophytes.

It is not difficult to see that these cannot be subject to death in the same degree as higher animals are.

(1) In the first place, being single cells, without any "body," they are able to sustain the equation between waste and repair for an indefinitely long period. It is conceivable that some of the simplest may

ABSENCE OF DEATH

have been living on since life began. They make good their waste by continuous and perfect repair. This has been summed up in the epigram that death was the price paid for a body.

(2) In the second place, it is a well-known fact that among multicellular organisms reproduction is attended with loss of life. One of the simplest—an Orthonectid—dies in giving birth, and the same is true of some worms. Death follows close on the heels of reproduction in the case of animals so different as may-flies, butterflies and lampreys. Everyone knows that flowering and fruiting exhaust the energies of annual plants. In the very morning of life immortality was pawned for love.

In the Protozoa and Protophytes, however, where the distinction between "body" and reproductive elements has not been differentiated, reproduction is a simpler, less expensive process. The Amoeba divides into two, only a metaphysical individuality is lost. There is as little death as when two cells fuse into one, another familiar reproductive phenomenon. Similarly, with spore-formation and budding, we cannot speak of death when there is nothing—not even ashes—left to bury. More prosaically it may be said that the conception of natural death which applies to the multicellular organisms does not apply in the same degree to those which are unicellular.

Maupas has indeed pointed out that an isolated family of Infusorians, all descended by asexual multiplication from one cell, and therefore not coupling or conjugating with one another, will, after a certain number of generations, come to extinction. But this isolation is hardly a natural condition. Nor, of course, does he deny the violent death of Protozoa.

(3) Thirdly, it is worthy of note that at least many Protozoa are not subject to death from bacterial infection to the same degree as higher animals. The Amoeba, for instance, seems but little perturbed by the presence of various virulent microbes. It engulfs them and digests them, as the phagocytes of higher animals do when in vigorous health, or when the odds against them are not too strong.

J. ARTHUR THOMSON, *The Science of Life.*

ALTERNATION OF GENERATIONS

We have seen that plants and animals produce asexually, and that they also produce sexually, that is to say, they develop into males and females which produce male and female reproductive cells.

If we consider the HYDROZOA, which usually form colonies of incrusting animals made up of units called *zooids* rooted to rocks in the sea, we find that they produce *asexually*, by budding, forms called *jelly-fish*, or *medusae*, which are quite unlike the parent that produced them. These jelly-fish develop male and female reproductive organs (testes and ovaries) and they reproduce, in their turn, sexually. They break off from the parent hydroid and swim away. The

FIG. 58. *Bougainvillea fructuosa*, magnified about 12. From Allman. *A*. The fixed hydroid form with numerous hydroid polyps which produce by budding (asexually) the jelly-fishes or medusae. These are seen in various stages of development. *B*. The free-swimming sexual medusa broken away from *A*, will produce by ova and spermatozoa (sexually) the hydroid form.

ALTERNATION OF GENERATIONS

fertilized egg of a jelly-fish does not produce a jelly-fish but a hydroid. Thus we have here an alternation of generations —a sexual generation (the jelly-fish) producing sexually, by ova and spermatozoa, an asexual generation (the hydroid). This hydroid in its turn produces asexually, by budding, a jelly-fish again.

Such alternation of generations is generally associated with the necessity of spreading the species, and it is in the animal kingdom as a rule confined to animals in which one generation is fixed, *sessile*, or to animals in which one generation is parasitic, such as the liver-fluke and the tape-worm. In the latter class the chance of finding the right host is very small. Hence everything is done to increase the number of reproductive cells: in certain tape-worms, for instance, the fertilized egg produces a larva which by internal, asexual, budding will produce dozens more larvae, and each of these by internal budding may again produce scores of offspring, all of whom may, if lucky, grow up into adult sexual tape-worms. Then again, the product of a single fertilized egg in a liver-fluke ends in hundreds of thousands of larvae, and the chance of one of them hitting the right host is thus considerably increased.

But, as we have seen, alternation of generations in the animal world does not reach very high up the scale. It is common in certain water worms; but it does not exist in the higher crustacea, or in mollusca; neither is it found amongst spiders or true vertebrates, nor in many of the other smaller and less conspicuous groups. It attains its highest point in the ASCIDIANS, where certain free-swimming forms, such as *Salpa*, have an alternation of generations. The Ascidians come at the very bottom of the vertebrate stem.

In plants, however, alternation of generations can be traced, although greatly masked, up to the very highest forms of vascular flowering plants.

In the simplest of the brown seaweeds we find plants which consist of branched cellular threads—*filaments*; these reproduce by both the methods which have been described; sometimes they form in their cells zoospores which germinate directly into new plants and sometimes they produce gametes which normally fuse before they can reproduce the filament. It is known that the proportion of asexual to sexual reproductive bodies changes with the time of the year. In the spring the majority of the plants produce gametes, in the autumn they form zoospores. We can imagine that in the course of evolution plants appeared in which this seasonal differentiation was firmly impressed in the life of the individual.

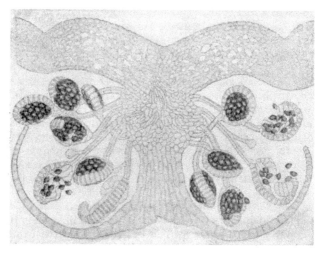

FIG. 59. Vertical section through a fertile leaf of a Fern, *Nephrodium*. Upon it are seated the numerous stalked sporangia, each containing many dark-walled spores. The whole is covered over by the umbrella-like indusium. After Krug.

In many of the higher algae of this group there is a very fixed alternation of plants forming zoospores which on germination produce plants which form gametes only; in some the two kinds of plants are alike to look at, in others they are quite different; thus in the large *kelp* (*Laminaria*) the plant which is so common produces spores, and these grow into minute filaments about a tenth of a millimetre in length. On these small threads the sexual cells are produced; after their fusion the product grows into the *Laminaria* plant, which may reach

SPOROPHYTES AND GAMETOPHYTES

two metres in length. When we come to deal with land vegetation, this alternation of generations in some more or less modified form is the absolute rule of all the higher forms.

A bracken-fern affords a typical example in which such an alternation of generations can be easily and readily recognized. If the back of a leaf of a fern is examined, there will be found certain brownish powdery patches or stripes. These are composed of receptacles—*sporangia*—which contain spores. The spores are quite asexual, and when under certain conditions the receptacle becomes ruptured, these minute cells are easily blown about and scattered throughout the neighbourhood. Since the fern, as one sees it, bears spores, it is called a *Sporophyte*. If the history of the spore is followed, it will be found that it comes to rest in very damp soil and gradually develops into a little pad or heart-shaped green plant known as the *prothallus*. The prothallus produces ova and antherozoids. Since the cells which fuse are called gametes, the prothallus which bears them is called a *Gametophyte*.

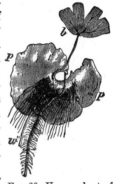

FIG. 60. Young plant of Maidenhair Fern still attached to prothallus (p); w'' roots; b, first leaf. After Sachs.

FIG. 61. Sexual organs of a Fern (*Polypodium vulgare*). I. Ripe female organ in section; p, prothallus; o, egg or ovum. II. A, the male organ which produces antherozoids. B, single antherozoid with its cilia for swimming, much enlarged.

The antherozoids require water to swim in so that they may reach and penetrate the ovum. Hence the prothallus must live in a situation where at least from time to time it has a watery coating.

REPRODUCTION

In certain cases the prothallus is unisexual, that is to say, unlike the prothallus of the bracken-fern it has either male or female gametes but not both. In that case the spores which give rise to the prothallus may differ in size. The larger, or *megaspores*, grow into the female prothallus; the smaller, or *microspores*, into the male prothallus. In the fern, the spore that gives rise to the gametophyte separates away from the body of the sporophyte (*i.e.* the flourishing green plant known as the fern). But as certain medusae or jelly-fish will not separate away from their hydroid stock, so the gametophyte and the sporophyte may continue in contact all their life. In the moss, the gametophyte is the dominant organism; it is in fact the moss plant with its tiny leaves, etc., but it embraces the sporophyte or *capsule* which remains embedded in the body of the former. Perhaps it is in mosses that the gametophyte reaches its best, for in the higher plants the gametophyte totally fails to make good. It is dominated and surpassed by the sporophyte, as is, of course, very obvious in the fern. This is still more obvious in the seed-bearing flowering plants. The pollen sacs in the anther which produces the pollen grain are the microsporangia, and the pollen represents the microspores. The nucellus in the female flower is the megasporangium and the embryo-sac is the megaspore which after fertilization will give rise to the new plant, the sporophyte.

Mosses and ferns at the most critical part of their life, that is when the fertilization of the egg is taking place, must have moisture; the antherozoids cannot move in the air—they must swim. In a way, these plants are the amphibia of the vegetable world; just as the frog has to descend into the pond to spawn, or a land crab into the sea, so mosses and ferns must have a certain amount of water for their male cells to swim towards the ova. A modern poet has written:

> Magnificent out of the dust we came
> And abject from the Spheres.

But I do not think the modern poet has made any deep study of the subject. The general consensus of opinion is that the land flora and fauna came from the waters.

WATER AND LAND PLANTS

Coming on shore has perhaps had a more profound effect on evolution than any other factor. It is only when land was reached that the higher plants and animals arose. But a change from an aqueous to a terrestrial existence had to be met by many modifications. Animals and plants are, as a rule, of about the same specific gravity as water. Hence they are uplifted, buoyed up, and do not require to make great efforts to move about in that medium. Certain animals move, like birds and bats and many insects in the air, in space of three dimensions, whereas animals that live on the earth are confined to one plane. Land plants, on the other hand, must develop certain organs for fixation. Their tissues must be strengthened so that they can support their own weight and resist the winds and storms which harass the surface of the earth. Then again, their watery protoplasm must be protected by such substances as cork from undue evaporation and their most external cells must develop a cuticle or they would dry up. But perhaps the most essential feature of terrestrial vegetation is the fact that the egg is protected within the parent plant.

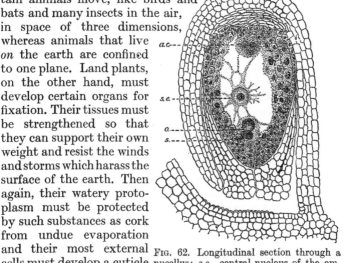

FIG. 62. Longitudinal section through a nucellus; *s.e.*, central nucleus of the embryo-sac; *o*, egg; *s* and *a.c.*, other cells of the embryo-sac. After Darwin.

As we have seen, the sporophyte in the land plant is the dominant partner. So long as the gametophyte is small and insignificant, as it is in the fern, it is in danger; but when gradually, step by step, the gametophyte became embedded and retained within the body of the sporophyte (as seems to have happened in the evolutionary history of the flowering plants) greater security was achieved. In those cases where the prothallus is unisexual and free living, the male antherozoid having fertilized an egg, the male gametophyte becomes useless and dies; but the female prothallus has not only to produce the egg, but to nourish it after it has been fertilized. For a time in certain ferns and many fossil plants the megaspore was dependent on its own resources. Safety was found in the higher seed-bearing plants by retaining the megaspore with its resultant gametophyte (prothallus) within the body of the sporophyte.

The *nucellus* of the plant corresponds with the megasporangium. Within it is the *embryo-sac*, which grows by feeding on the surrounding cells, and absorbing their contents. Within the embryo-sac are cells, one probably representing the ovum or egg, and the others relics of the primitive prothallus. In flowering plants therefore the prothallus is hidden away inside the tissues of the parent, just as in a mammal the embryo is hidden away inside the body of the mother. The pollen-sacs of a flower, borne at the free end of the *stamens*, represent the microsporangia. These give rise to pollen-grains, which represent the microspores. Pollen-grains

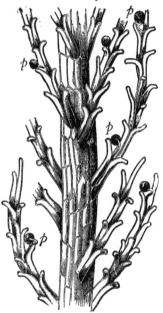

FIG. 63. Pollen-grains (*p*) germinating on the stigma of a Grass. After Kerner.

are blown or carried in some way to the *stigma* of the female organs and emit a long pollen tube. The pollen-grain contains two cells. One is usually larger than the other. It is called the *vegetative-cell*, and probably represents the vegetative part of the prothallus. The other and smaller cell divides into two male gametes, each of which represents an antherozoid of the original prothallus. One marked advance was gained when

NUMBER OF SEEDS

the antherozoid ceased to swim and pollen-grains could be conveyed through the air from flower to flower by wind or by insect agency.

The fact that the prothallus has been maintained in this highly modified form in the higher seed-bearing plants was for a long time not recognized. But as the result of a series of comparative studies there seems no doubt that this is so. The seed which contains the embryo plant is a direct development of the ovule. It is detachable and contains an embryo which will grow into a new plant. It also contains food, generally in the form of starch and protein, for the embryo to live on.

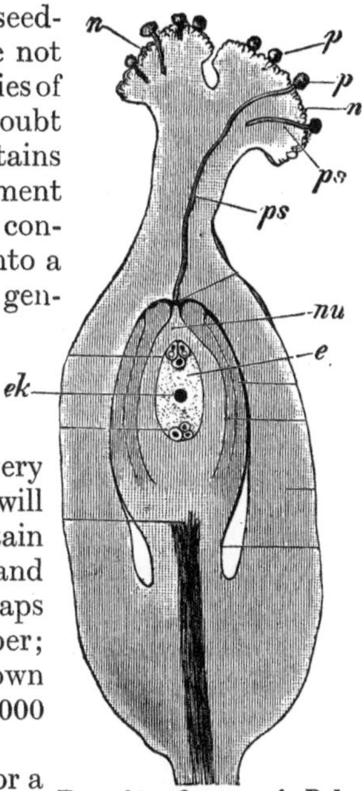

Number of Offspring

The number of seeds is often very great. During one season the radish will produce 12,000 seeds, the plantain 14,000, the shepherd's purse 64,000 and the tobacco plant 360,000. But perhaps the orchids show the greatest number; a single plant of a certain genus known as *Acropera* is said to produce 74,000,000 seeds in one breeding season.

Whenever you look at a garden or a wood or a field, it is seed-bearing plants that you see. The others, non-seed-bearing plants, are insignificant in comparison with this dominant group; part of their success is undoubtedly due to the fact that they protect the female reproductive cell within their own body, nourishing it and cherishing it in every way.

FIG. 64. Ovary of *Polygonum convolvulus* during fertilization; *nu*, nucellus; *e*, embryo-sac; *ek*, central nucleus of embryo-sac; *n*, stigma; *p*, pollen-grains; *ps*, pollen-tubes. Magnified 48. After Strasburger.

The plant as we see it is asexual; protected by numerous coats and layers of cells is the embryo-sac, which "in a manner of speaking," as John Silver used to say, is a parasitic prothallus inside the asexual body of the sporophyte.

176 REPRODUCTION

With the exception of the great group of insects, spiders and the higher vertebrates and a few others, most of the dominant classes of animals live in or on the water, and amongst them as a rule the females shed their eggs into the surrounding aqueous medium, and the males standing by shed their spermatozoa near the eggs. In many cases the eggs and the spermatozoa exist in simply incredible numbers. Amongst the fishes the female turbot will produce annually 9,000,000 eggs, the cod 6,000,000, the flounder 1,400,000, the sole 570,000, the haddock 450,000, the plaice 300,000, whilst the herring has to be content with comparatively few, its meagre total reaching only 31,000. The spermatozoa must be produced in numbers that are even far greater, for it is rare to find an unfertilized egg in the sea; and yet many spermatozoa must miss their goal. The reflection that if the stock of cod remains about constant—as indeed it does—only two out of 6,000,000 eggs attain maturity, almost paralyses imagination as to the destructive forces at work.

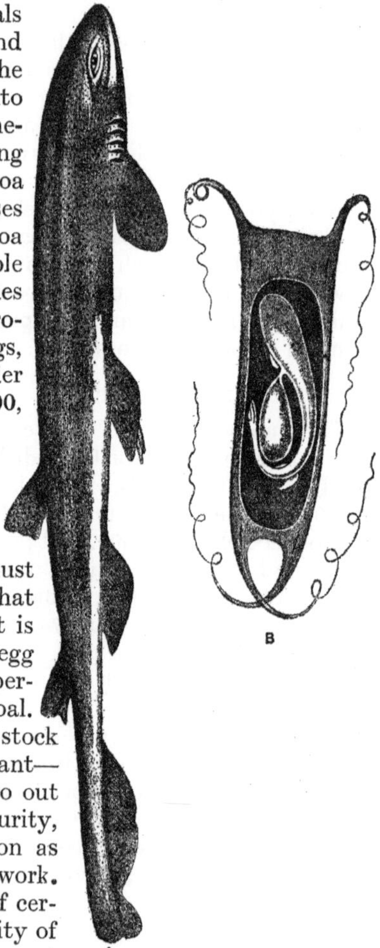

The amazing fecundity of certain animals and the paucity of offspring of others is an apparent paradox. Whilst a turbot will produce 9,000,000 eggs, of which presumably only two survive—or

FIG. 65. *A*. A Dogfish, *Scyllium canicula* Reduced. From Day. *B*. Egg-case opened to show young embryo with yolk-sac.

PARENTAL CARE

turbots would be much cheaper than they are—the dogfish or skate produce only a dozen or so. But in spite of this fact the dogfish is much more abundant and much more widely distributed than the turbot. Darwin tells us that "no fallacy is more common amongst naturalists, than that the numbers of an individual species depend on its power of propagation." When you have an enormous number of eggs laid they must take their chance, and so great is the struggle for existence that the chance is, at best, a very poor one. A fish producing millions of eggs can exert little maternal solicitude for her numerous brood: "a mother's tender care" is necessarily lacking. But when the fertility is small and the number of offspring few, there is always protection for the eggs and larvae from enemies.

On the one hand enormous numbers of eggs are produced only to be destroyed, while a few survive to maintain the species; on the other hand few eggs and offspring are brought into the world and all kinds of devices are elaborated for the protection of these.

Parental care can only be exercised when the number of eggs is small; and it is this maternal or paternal solicitude which enables the species to persist, although the number of eggs may be comparatively very few. A good instance of this is the well-known fish the stickleback, which occurs both in the sea and in fresh water. The male constructs a nest of weeds and twigs fastened together by threads secreted from his kidneys. The nest is a spherical structure with a small hole left on one side for entrance. When it is completed the male seeks out the female and conducts her with many caresses to the nest, introducing her through the aperture. She passes in and lays two or three eggs, and leaves the nest by boring a fresh hole at the side opposite to that by which she entered. The nest has now two doors through which a current of water flows. Next day the male seeks out the same or another female and brings her to the nest, where she lays one or two more eggs. This is repeated till the nest contains a fair number of ova. The male will then watch for a month over the nest, defending it against all invaders, especially against his temporary wives who show a most unnatural

178   REPRODUCTION

instinct to get at the eggs. Once they are hatched, the parental care is at an end, and the young fish fend for themselves.

In many other fish, such as the sea-horse, the lump-sucker, etc., and certain frogs, the males guard and tend the eggs of their mates. In fact, amongst the fauna of the sea-shore paternal care is as common, if not more common, than maternal. But in land-vertebrates it is the mother as a rule that rears, tends and fosters her offspring.

A very large number of water animals and plants have their eggs fertilized whilst floating freely in the sea or fresh water; the antherozoids of the latter and the spermatozoa of the former are adapted for swimming through an aqueous medium.

Life originated in the sea, and certain considerable groups of animals have never left it. In spite of deserters, the marine

FIG. 66. *Amphioxus lanceolatus* from the left side, about twice natural size
After Lankester.

fauna is still far richer and more varied than that of the fresh water or of the earth. In fresh water and on the earth there are none of those flinty little Radiolarians which form so large a part of the sea bottom, nor are there calcareous Foraminifera in our streams and lakes. Chalky and horny sponges exist only in the sea, and many of the larger sub-groups of the Coelenterata do not penetrate fresh waters, though occasionally a jelly-fish is found. Curious worm-like animals known as GEPHYREA and CHAETOGNATHA are found only in the ocean. Many large groups of the MOLLUSCA, prominent amongst which are the cuttle-fishes and the PTEROPODA upon which the young whales so largely live, are exclusively marine. No specimens of the groups ECHINODERMATA, starfishes, sea-urchins, sea-lilies, occur outside the sea, and one geologically extremely old group, the BRACHIOPODA, are all purely marine in their habitat, and so are the little forerunners of the VERTEBRATA, the ASCIDIANS and a primitive little fish-like form known as *Amphioxus*. The

LAND AND SEA ANIMALS

fresh waters of our globe have one group of fishes, the lung-fishes, DIPNOI, which do not occur in the ocean, and there are no marine Amphibia. The air-breathing insects and spiders are, with very few exceptions, never marine. There is one truly sea-going insect, *Halobates*, a species of bug which lives in the Sargasso Sea. Certain animals that have taken to land have in time returned to their ancestral home. Amongst these may be reckoned, on the sea shore, a few spiders, mites and insects, the very poisonous sea-snakes, and probably the dolphins. The great bulk of the crustacea are marine and fresh-water, but mainly marine. They, like so many other large groups, have never taken to the air. When the authors of the *Anti-Jacobin* so passionately exclaim:

> Ah! who has seen the mailéd lobster rise,
> Clap her broad wings, and soaring claim the skies?

the answer, in the language of those curious mammals, the politicians, is "in the negative."

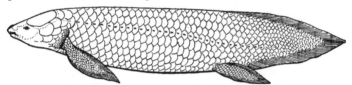

FIG. 67. A Lung-fish, *Ceratodus forsteri*. After E. S. Goodrich.

Great expanses of inland waters are salt and some contain a number of marine genera; thus the Caspian, though less salt than the ocean, harbours the seal, the salmon and the herring. But such lakes as the Dead Sea and the Great Salt Lake of Utah have become so concentrated as to render almost all life impossible.

> The saline Lakes grow salter by degrees
> 'Till pickled salmon swim the astounded seas.

ORIGIN OF LIFE FROM THE SEA

One of the greatest adventures of the organic world must have been when, as Solomon in his "Wisdom" tells us,

The things that before swam in the water, now went upon the ground.

and among Vertebrates it was the Amphibians that began it. At present Amphibia form but a small and inconspicuous Class of Vertebrates and they show no particular enterprise. But in Carboniferous times, when our coal-measures were being laid down, their ancestors must have shown a wonderful spirit of adventure. They are clearly derived from a fish-like stock, possibly from one of the lung-fishes, in which the swimming-bladder has been converted into a respiratory organ. Even at the present time certain Amphibian genera may be permanently aquatic, and with very few exceptions their larvae are always cradled in the water. Whatever their ancestry was in the sea, they must have come on land *via* the fresh water. All the lung-fishes live in rivers and the Amphibia live in fresh water or lay their eggs there, never in the sea. It is common enough among animals who have changed their habitat during their lifetime to return, when producing their young, to the place where they were born. Land-crabs, for instance, do this, and their eggs are always hatched in the sea, and conversely turtles who have taken to a marine habitat return to the land to lay their leathery eggs in the hot, sandy beaches. The amphibian tadpoles have many fish-like characters. They breathe by gills, they cannot move their tongue, they swim by the flexure of their body and tail, they have a two-chambered heart; but gradually all these change.

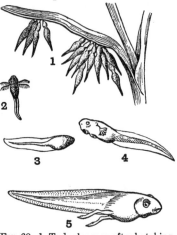

Fig. 68. 1. Tadpoles soon after hatching, clinging to water-weeds. 2. Tadpole with two pairs of external gills. 3 and 4. Tadpoles with operculum forward and forming. 5. Tadpole with well-developed hind legs.

The Amphibia are the lowest vertebrates which possess fingers and toes. They can grasp an object, they can even burrow and put food into their mouth. In the adult form they

COMING ON SHORE

have a motile tongue, which in the frog is fixed in front and not, like ours, behind, and can be shot out at a great distance to catch their insect food.

Again, amongst Amphibia, we for the first time come across a vertebrate voice. The serenading of the frog recorded by Aristophanes is produced by the passage of air being breathed out through vocal chords stretched tight in the larynx. In Aristophanes' play of the *Frogs* their voices, which are very primitive, are reproduced by the following line:

Βρεκεκεκὲξ κοὰξ κοάξ

Brekekekex, ko-ax, ko-ax,

In Rogers' translation of this comedy there is an interesting note, which I here reproduce:

The croak of a frog has been one of the best means of informing the modern world of the manner in which the ancient Greeks pronounced their beautiful language. The frogs of the nineteenth century have probably been faithful to the pronunciation of their race in former times; and, as we listen in the still night to their curious music, it is exactly as if one set of them, perhaps the tenors, the gentlemen of the choir, kept singing "Brekekekex," while the softer wooing of the ladies is uttered always as "Koax, koax, koax."

Coming on land meant a diversity of habits, for the amphibian is primitively an aquatic animal. The Amphibia do not lay a very large number of eggs; but they lay them always in the water. The spermatozoa are poured out over them and are not introduced into the body of the female. The same is true of lobsters and cray-fish and certain other crustacea. The next two groups of Vertebrata, the Reptilia and the Aves, produce comparatively few eggs and they are fertilized inside the body of the mother. They are large and provided with a great amount of yolk to serve as food for the embryo. They are often protected by being buried in the sand, or otherwise tucked away in skilfully constructed nests.

It is in the mammalia that the embryo develops within the body of the mother. Here the egg is fertilized and the number of the spermatozoa which pass into the mother's body is incredible. In some species there are estimated to be some 226 millions of spermatozoa; at other times 551 millions passing into the body of the mother in one act of pairing, and yet but one or two ova are fertilized. Both these figures and those of the fish eggs mentioned above seem to indicate an appalling waste, yet we must suppose these excessive numbers to be necessary in order to ensure a sufficiency of fertilized egg cells.

There are, however, other methods of fertilization. As a rule the spermatozoa follow in the body of the animal a definite channel; but there are worm-like animals, TURBELLARIA, into whose body they are introduced through the skin at any spot. They then have to make their way through the tissues of the body, the muscles, etc., until they reach the eggs. A further modification of this method of fertilization is that in some flat-worms, leeches, and even an animal as high up as *Peripatus*, packets, called *spermatophores*, of spermatozoa are made up. These packets are implanted in the animal like the arrows in St Sebastian's body. The tip of the implanted spermatophore is then dissolved and the spermatozoa stream out and force their way through the muscular and connective-tissues and other fabrics of the body until they reach the eggs.

Packets of spermatozoa or spermatophores are found in other classes, for instance, in the snail, where a hardened mass of mucus whose edges are rolled round so as to form a groove holds the spermatozoa as in a case. This case or spermatophore is deposited by the snail acting as the male, for the snail is both male and female, in the genital orifice of another snail. In a few days the case of the spermatophore is dissolved and the contained spermatozoa are set free ready to fertilize the ova. In the *Sepia* also we find cylindrical spermatophores which are not inserted into the ordinary genital orifice, but are deposited on the skin of the female, near the head. When the eggs are laid, the spermatozoa flow out and fertilize them in the water.

THE EGG

As a rule in both plants and animals the egg is spherical; it may be, and often is, microscopic; but when a large amount of yolk is stored up in it, it may reach gigantic proportions.

EGGS AND ANTHEROZOIDS

The eggs of the birds before they are released from the ovary are amongst the largest cells we know. The egg of the ostrich is from 150–155 mm. long and 110–130 mm. wide. It weighs 1440 gms. and is equal to some 24 fowls' eggs. But the extinct *Aepyornis*, of Madagascar, several of whose eggs are known, surpasses even this, being about 340 mm. long and 220 mm. wide. It has a capacity of 3·7 litres, and equals at least twelve ostrich eggs or 288 average domestic fowls' eggs. It is 500,000 times bigger than the small egg of the humming-bird. The curious New Zealand kiwi, *Apteryx*, has an egg quite out of proportion to the size of the bird that lays it. The egg weighs almost a quarter of the body-weight. Of course it must be remembered that these eggs consist almost entirely of yolk, but the amount of metabolism that must go on in the bird's body to produce such an ovum must be tremendous. The actual protoplasmic egg is minute.

Occasionally ova are amoeboid. This is true, as we have seen, of the *Hydra*. Very often the egg is oval or sausage-shaped. This is frequently the case in insects, where the egg case may be a structure of great symmetry and beauty, resembling in some cases seeds and in others little chalices with a cross on the top. Sometimes the egg case has a special aperture or *micropyle* for the admission of the spermatozoa. More often the spermatozoa may penetrate at any part of the surface and, as a rule, when once a spermatozoon has penetrated, a membrane is formed which prevents the access of others; but this is by no means a universal rule and some eggs, such as those of the dogfish, receive many spermatozoa.

ANTHEROZOIDS AND SPERMATOZOA

As a rule the male gametes of plants move by means of two flagella. Some antherozoids have regular tufts or bunches of cilia, as in the adder's-tongue and other ferns. The two flagella may arise at one end of an oval body or they may emerge from the side, one stretching forward and one backward. Antherozoids have been modified in the flowering plants into two naked cells which appear in the pollen-tube and one of which fuses with a similar naked cell in the

REPRODUCTION

embryo-sac of the ovule. Like the spermatozoa of animals, they are microscopic. In animals each spermatozoon consists of a head which may be elongated, rounded, oval or corkscrew-shaped; it terminates in a long tail or flagellum. The whole of the nuclear matter of the cell is confined to the head, and when once the spermatozoon has penetrated the egg the tail is absorbed. As a rule they are extremely minute, in man only 0·05 mm. in length; but in certain small crustacea, known as OSTRACODA, they attain gigantic proportions. To take one example, in a genus known as *Pontecypris* they are from 5–7 mm. in length—five to ten times as long as the whole body of the animal. Certain members of the CRUSTACEA have flagellate spermatozoa, *i.e.* with tails, and this is true of the higher orders of ARTHROPODA, *i.e.* the insects, spiders, etc. These flagellae are the only structures of the kind found in the group, and in many, such as the crab and lobster, the spermatozoa are star-shaped and have to be placed in contact with the ova, as they have little power of locomotion.

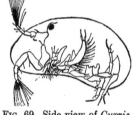

FIG. 69. Side view of *Cypris candida*, a typical Ostracoda. After Zenker.

The other large Class of animals whose spermatozoa are devoid of flagella are the round-worms, or NEMATODA. Here they are described as hat-shaped—a description which, considering the present fashions, does not impress a very clear image on the mind. The protoplasmic part of these spermatozoa is amoeboid.

Some males produce two different kinds of spermatozoa, larger and smaller; the latter are regarded as the only functional forms. The spermatozoa may obtain access to the ova after the latter have been laid, as is the case in many aquatic invertebrates and in most fish, or just as they are being laid, as occurs in lobsters, crabs, and in frogs and toads, or within the body of the mother. This is the case in many worms, leeches, insects, many molluscs and in all the higher vertebrates, reptiles, birds and mammals.

HERMAPHRODITISM

Most commonly ova and spermatozoa are produced by different sexes, females and males. This is, however, far less common in plants than in animals, for a considerable number of animals and most plants are *hermaphrodite*, that is to say, they have both male and female organs in the same body. Common examples of these are the majority of plants—which are then called *monoecious*—the sponges, coelenterates, leeches, earth-worms, snails and some crustacea and molluscs.

The hermaphroditism of certain forms is a permanent feature; but in other animals it may be a matter of time: in a mollusc known as the slipper-limpet, *Crepidula*, which is so destructive to our oyster beds at Colchester, the growing animal is first of all neuter, then it becomes a male, then it changes into an hermaphrodite, and lastly it becomes a female.

Similar changes take place in certain hermaphrodite NEMATODA, in the TURBELLARIA, and in the parasitic CRUSTACEA known as the RHIZOCEPHALA, spermatozoa developing in the ovary in increasing numbers in successive generations, until the animals become completely monoecious. In the Vertebrata hermaphroditism is abnormal. Hermaphrodites generally pair with another hermaphrodite, the spermatozoa of specimen *A* fertilizing the ova of *B*, and the spermatozoa of *B* fertilizing the ova of *A*, so that there is a cross-fertilization. But there are other cases, especially among the parasites, where the spermatozoa of an animal fertilize its own eggs. Both flukes and tape-worms are at times self-fertilizing.

The dramatist Sheridan has told us that even "an oyster can be crossed in love"; but the physiology of the reproduction of that mollusc was not well understood in the days of that gifted dramatist. It seems to be clear now that a young oyster is a male, that after it has discharged its spermatozoa it may be changed into a female at the age of one year, and after it has spawned it may again revert to the male condition. The changes of sex are repeated, and as far as one knows they may be due to external causes, such as temperature. As Stephen Paget tells us, "Oysters in their way are quite as wonderful as poets, saints and men of science."

Somewhat similar conditions occur in the free-living NEMATODA. The nematodes living in the soil present three different forms of sexual condition:

(a) separate sexes,
(b) hermaphroditism,
(c) parthenogenesis.

It is believed that the hermaphrodite species have been developed from bi-sexual species by the development of spermatozoa in the female and the suppression of the male sex. In several species males exist, often in considerable numbers, which are completely developed and produce spermatozoa, but yet do not function. The hermaphrodite nematodes are all self-fertilizing, as is the case with TURBELLARIA and with flukes and tape-worms. Hermaphroditism in the round-worms is, however, incipient, and not yet fully established, for the number of spermatozoa provided are quite insufficient to fertilize the eggs that are produced. When the spermatozoa are exhausted, the unfertilized eggs go on being laid, but rapidly disintegrate.

Hermaphroditism can be induced. When the hermit-crab becomes infected by a certain parasite known as *Peltogaster*, ova make their appearance in the testes of the male crab and its secondary sexual characters put on a female appearance; and the same is true of another crab, *Inachus*, when infested by a degenerate crustacean parasite, known as *Sacculina*. Then there is a certain marine annelid, known as *Bonellia*, whose larvae after a free-swimming stage become attached to the sea bottom and develop into females or become degenerate males. By releasing the larvae from their support before their sex has become established and forcing them to lead an independent life, it was found possible to produce grades of individuals in which the degree of maleness or femaleness depended upon the duration of the time they had been attached before being released.

PARTHENOGENESIS

Parthenogenesis is the development of a true egg without being fertilized by a spermatozoon. It can be produced artificially; but its normal occurrence in nature is very common.

PARTHENOGENESIS

Parthenogenesis is common in plants. There is a lowly fresh-water alga called *Chara crinita* whose egg develops without union with an antherozoid. In fact, in Europe, where it flourishes, only female plants of the species are known, so there is no chance of fertilization. Many flowering plants produce parthenogenetically. This is perhaps especially true of cultivated plants such as the Compositae (daisies, etc.), members of the rose family, buttercups, thymes, stinging-nettles and many other plants. In these the naked ovum in the embryo-sac develops into the new plant without being fertilized by what corresponds to the antherozoid in the pollen-tube.

After a queen-bee has been paired high up in the air she returns to the hive with no less than 200,000,000 spermatozoa in her body, a supply equal even to her prodigious fecundity, for she will lay 2500 to 3000 eggs every twenty-four hours, during three or four years. The eggs which are destined to produce the queen and the workers are fertilized, but, by some marvellous arrangement which is not yet fully understood, the spermatozoa are kept back from the eggs destined to form the males or drones, the males being born parthenogenetically.

Amongst the ROTIFERA parthenogenesis is very common, especially during the summer months; fertilized eggs appear in the autumn but they are known as winter or resting eggs. They are capable of resisting adverse circumstances and carry on the life of the race during the winter. But in many ROTIFERA no male has ever been found and presumably these forms are entirely parthenogenetic.

Parthenogenesis also occurs in the life-history of many of the lower CRUSTACEA. Both in the PHYLLOPODA and in the OSTRACODA unfertilized eggs are produced. Here again in some cases males are totally unknown, but in others the male does appear towards the autumn; the eggs it then fertilizes pass into a resting stage and are protected by a capsule formed by the skeleton of the mother. In many cases males are very rare, sometimes only one per cent. of the total stock. In many insects a similar state of things prevails and there may be here, as among the lower CRUSTACEA, many parthenogenetic generations before the male reappears.

188   REPRODUCTION

In insects there are still further complications. In the case of the plant lice *Aphis* we have a winter egg which produces the female in the spring. This will continue to produce generation after generation parthenogenetically. These are wingless; but quite unexpectedly and suddenly a generation will appear which has wings, and these wings undoubtedly aid in spreading and scattering these most devastating enemies to plant life. In an allied form of *Phylloxera*, which has done so much to destroy the vines in France, we even have three different kinds of female instead of two (the wingless and the winged), as in *Aphis*; these females reproduce parthenogenetically before the male at last turns up.

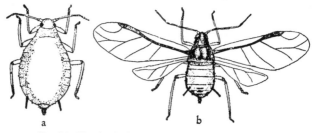

FIG. 70. The Apple Aphis, *Aphis pomi*, virgin females. *a*, wingless; *b*, winged. Magnified 20. From Carpenter.

Plant lice increase at a perfectly appalling rate. I once heard a distinguished lecturer in America tell his class that if all the young of a plant louse survived, and all the young of the survivors survived, and so on, in the course of one year there would be a column of plant lice with a cross-section of one square mile advancing into space with the velocity of light!

In somewhat higher insects, such as the saw-flies (TENTHREDINIDAE), the gall-flies (CYNIPIDAE), and the scale-insects or mealy-bugs (COCCIDAE), no male has ever been found. It is a curious fact that in some cases of parthenogenesis amongst insects the product is always of one sex, some clutches of eggs producing always females, the other always males.

There is no doubt that amongst the classes mentioned in the foregoing paragraphs, the females seem to get on perfectly well without any male, and it may here be stated that until we reach the Vertebrata, and even in some of them, the male is nearly always smaller and feebler than the female. In many cases, as in the honey-bee, the male dies the moment after pairing. Having once paired, his functions are at an end, whilst the female has to carry on the race. In some cases, such as in that curious marine worm, *Bonellia*, the male is reduced to a mere parasite in the body of the female. I used to try to conceal these facts in the time of the suffragette agitation, but sooner or later they were bound to come out.

The alternation of parthenogenetic eggs with fertilized eggs is another example of the alternation of two different generations; but it is not quite on a level with the cases mentioned above: for here the asexual generation is in reality an unfertilized egg, whereas in the case of the lower animals, such as the HYDROZOA or the flukes, the alternation is between a vegetatively produced sexual generation and a sexually produced vegetative generation. In some cases we have seen that in both kinds of alternation of generation the stages may not be strictly alternate. If S represents the sexual generation and A the asexual, you may have many A's before the sexual turns up again, such as $S\text{-}AAA\text{-}S\text{-}AAA\text{-}S$.

Parthenogenesis can be induced by artificial means in eggs which are normally sexual. Certain changes in the food or environment or certain stimuli will induce or quicken reproduction. The members of a pure culture of *Paramoecium* descended from a single parent may live in many hundreds of generations. A long period of reproduction by simple fission is followed by a period when conjugation is common. This starts a new cycle and is physiologically comparable to the period of fertilization in the higher animals. If such conjugation be prevented the individuals suffer senile decay and die. But artificial stimuli can tide over this period of growing old and of decay. In a certain culture which lasted nearly two years the *Paramoecia* periodically began to fail every six months, and they would have disappeared entirely had it not been for the application of such stimulants as beef-tea, alcohol, extracts from the brain and pancreas or even a shower of rain. A change of food or environment produced a rejuvenescence, just as conjugation does.

Other observers have been able to maintain a single race of *Paramoecia* for over five years and carry it through three thousand generations by simple fission without conjugation having occurred. This was accomplished by continually altering the character of their food and as far as possible imitating the conditions of pond-life. Another culture which is, I believe, still "going strong" was started some fifteen or more years ago.

One of the ductless glands of the Mammalia, the *pituitary body*, situate at the base of the brain, is associated with sexual changes, and by administering extracts of this gland to *Paramoecia* reproduction was markedly increased. This is a remarkable fact, for the *Paramoecium* is as widely removed from the mammal as any animal could well be. The secretions of the pituitary body which pass into the blood also have a great effect on the growth of the body.

Artificial parthenogenesis has been known since the end of the last century. Eggs of the silk-worm moth and of the marine worm, *Chaetopterus*, can be induced to show the early signs of fertilization by the addition of certain chemicals. By concentrating the amount of salts it was possible to induce the unfertilized eggs of the sea-urchin to develop as far as the larval stage, termed the *Pluteus*. The rate of development was slower than when the ovum had been fertilized by a spermatozoon and there were other irregularities; but still the larvae were there. Similar experiments succeeded more or less with the eggs of certain limpets and with those of the lamprey and of the frog.

Unfertilized eggs may be induced to develop not only by the application of certain salts and the variation of the media in which they live, but by electrical and mechanical stimuli.

Gently brushing with a camel's-hair brush or applying electric discharges to the egg, even by submitting the ovum to pin-pricks, will start the unfertilized egg segmenting, but, as a rule, it does not go very far. Occasionally larvae and even adults have been reared as the result of these mechanical processes. A young sea-urchin has been raised from an unfertilized egg, and by pricking frogs' eggs with a platinum needle, which was sometimes dipped in salt or in the blood of the mother, a tadpole and, on very rare occasions, minute frogs have been produced. In these cases the nuclei are always smaller than normal and contain only one-half the normal number of certain constituents of the nucleus, the chromosomes. Another curious fact is that by making sea water markedly alkaline, the spermatozoa of more than one species of star-fish were induced to fertilize the eggs of sea-urchins.

Normal sexual reproduction involves the fusion of a microscopic spermatozoon with an ovum which in many cases is also microscopic. These exceedingly minute cells are the carriers of heredity and within their small bodies are contained the future characteristics, mental, moral and physical, of the resulting offspring. It is the fusion of these two cells which produce not only the features which are common to both parents, but also a combination of features due to the mixture of their protoplasms and nuclei. Offspring resemble their parents to a greater or less extent—"Do men gather grapes of thorns, or figs of thistles?" But they also vary from their parents to a greater or less extent and it is this variation, which may be fixed and inherited "from one generation to another," which leads to the establishment of new types in the living world.

Old Age and Death

Life is a cycle, beginning with an egg and coming round in time again to an egg.

The fact that physiologically life is a wheel, a circle, a cycle, is expressed by Sir Michael Foster in far better words than I can command:

When the animal kingdom is surveyed from a broad stand-point, it becomes obvious that the ovum, or its correlative spermatozoon, is the goal of an individual existence: that life is a cycle beginning in an ovum and coming round to an ovum again. The greater part of the actions which, looking from a near point of view at the higher animals alone, we are apt to consider as eminently the purposes for which animals come into existence, when viewed from the distant outlook whence the whole living world is surveyed, fade away into the likeness of the mere byplay of ovum-bearing organisms. The animal body is in reality a vehicle for ova; and after the life of the parent has become potentially renewed in the offspring, the body remains as a cast-off envelope whose future is but to die.

Once the plant or animal has secured the continuation of the species, its body is of little further use. It grows old and decays and in due course dies; and it is well that it should be so.

> The days of our years are threescore years and ten;
> And if by reason of strength they be fourscore years,
> Yet is their strength labour and sorrow;
> For it is soon cut off, and we fly away.
>
> Psalm xc. 10.

As the melancholy Jaques in *As You Like It* tells us:

> The sixth age shifts
> Into the lean and slipper'd pantaloon,
> With spectacles on nose and pouch on side,
> His youthful hose, well saved, a world too wide
> For his shrunk shank; and his big manly voice,
> Turning again toward childish treble, pipes
> And whistles in his sound. Last scene of all,
> That ends this strange eventful history,
> Is second childishness and mere oblivion,
> Sans teeth, sans eyes, sans taste, sans every thing.

The decay of the faculties is said to take place in inverse order to the acquirement of them. The average man when he reaches the age of forty-five has to have recourse to spectacles, and his vision decays as he grows older. In old age the hair and the teeth tend to disappear, the hearing is deadened. His body becomes smaller. He is physically weaker, and in old age we return to our earliest memories, forgetting the things of to-day and remembering best the things that have long since passed. In this weakened stage it is obvious that no fate could be more unhappy than immortality. When Eos begged Zeus that her lover Tithonus might live for ever, she made a profound mistake. What she ought to have asked for was that he should be endowed with immortal youth. This error of judgment cost her dear, for Tithonus shrivelled up into a hideous old man whom Eos kept shut up in a chamber. Finally his prayer was granted and he was changed into a grasshopper.

No fate could be more unhappy than immortality in this world.

> The woods decay, the woods decay and fall,
> The vapours weep their burthen to the ground,
> Man comes and tills the field and lies beneath,
> And after many a summer dies the swan.
> Me only cruel immortality
> Consumes:

IMMORTALITY

> I ask'd thee, "Give me immortality."
> Then didst thou grant mine asking with a smile,
> Like wealthy men who care not how they give.
> But thy strong Hours indignant work'd their wills,
> And beat me down and marr'd and wasted me,
> And tho' they could not end me, left me maimed
> To dwell in presence of immortal youth,
> Immortal age beside immortal youth,
> And all I was, in ashes.
>
> TENNYSON, *Tithonus*.

As Kim's Lama said about the cobra, "He is on the wheel as we are—life ascending or descending—very far from a deliverance."

Still, the wheel is advancing, not stationary. Things do get better. Even the Great War has shown us that

> Civlysation *doos* git forrid
> Sometimes upon a powder-cart.
>
> *The Bigelow Papers.*

Anyone who has lived through the last sixty years, as I have, must recognize that, in spite of temporary set-backs, the world is a better, a cleaner and a kindlier place to live in than it was in the middle of the last century. Mankind does advance, slowly if you like. He is said to have taken fifty thousand years to learn how to produce a polished arrow, or kelt, from a flaked one; but still in the end he did produce it, and it was a great advance in civilization.

The last two pages dealing with old age may to some produce the impression of melancholy and decay; but this should not be the case:

> Grow old along with me!
> The best is yet to be,
> The last of life, for which the first was made:
> Our times are in His hand
> Who saith, "A whole I planned,
> "Youth shows but half; trust God: see all nor be afraid!"
>
> R. BROWNING.

Further, it must be remembered that these pages deal only with the visible body, its cells and tissues. They leave out of account altogether other elements of our being which do not come within the scope of the present book. Perhaps one may be permitted to close the work by quoting the hackneyed, if hopeful, lines of Longfellow:

> "Dust thou art, to dust returnest"
> Was not spoken of the Soul.

INDEX

ACACIA, movement of, 120
Acanthocephala, 24
Acer platanoides, shoot of, 30 (fig. 8)
Acropera, number of seeds, 175
Adder's-tongue fern, 183
Adrenaline, 92
Aepyornis, egg of, 183
Aerobic organisms, 112
Air, composition of, 23
  food of plants, 24, 29
  essential to human life, 111
  in trout's air-bladder, 113
Aleurone-layer, Vitamin B in, 94, 95
Algae, 29, 61, 94, 151
  spores of, 162
  reproduction of, 164
Alimentary canal, 88, 89, 90, 91
  a means of breathing, 104
  absence of oxygen in, 113
Alligators, food of, 69
Alpine plants, vegetative reproduction of, 164
Alternation of generations, 167–9, 189
Alveoli, 111
Amaryllis, 54
Ambergris, value of, 76
Amino-acids, 29, 89
Ammonia, 21, 39, 40
Ammonium carbonate, 39
Amoeba, described, 9–13
  dissection of, 11–12
  distinguished from other organisms, 13
  killed by 98° F., 45
  and bacteria, 45, 59, 109
  movement of, 121, 123
  movement rate of, 122
  mode of reproduction, 161, 165
  immortal, 166
Amoebocytes, 121
Amoeboid movement, mechanism of, 122
  movement, rate of, 122
  ova, 183
Amphibia, 105, 179, 180, 181
Amphioxus lanceolatus, 178
Anabolism, 58
Anaerobes, 35, 38, 112, 113, 114

Annular sap vessels, 50
Ant-bear, food of, 74
Ant-eater, Australian, *Echidna*, 74, 109
  Tamandua, 76
Antedon, see Sea-lilies
Antherozoids, 15, 116, 119, 166
  rhythmic beat of flagella in, 134
Ants, specialized forms of, 154
Aphis, winter egg of, 188
Aphrodite (Sea-mouse), 108
Appetite, 96, 97
Apus, 107
Aragonite, 138
Arca, 107
Arcella, rise and fall of, 144
Armadillo, food of, 75
  armour of, 19
Arthropoda, excretion by, 5, 17
  movement of, 126
  spermatozoa of, 184
Artichoke, Jerusalem, 163, 164
Artiodactyla, 77, 78, 79
Ascaris mystax, 113
  A. megalocephala, 113
Ascidians, 5, 26, 58, 169, 178
Ass, wild, food of, 78
Assimilation by protoplasm, 4
Aster, rhythmic occurrence of, 136
Aurora Borealis, 37
Aves, generation of, 181
Axon of nerves, 121
Azotobacter, 38

BABINGTONIA ABOMINA
  BILIS, 162
Bacillus, derivation of name, 27
  B. subtilis, 27
  B. welchii, 113
  B. botulinus, 113
Bacteria, 26, 37
  contents of, in soil, 45
  in soil, 45
  rhythmic fluctuations in number, 45, 142
  food for protozoa, 59
  immortal, 166

INDEX

Bacterial infection, 167
Badger, 47
Bananas, 96
Barberry, stamens move, 119
Barnacle, 61
  larvae of, 151
Bats (*Cheiroptera*), 84
  "flying foxes," 145
Beaver, 81
Bee, queen, fertilization of, 187
Beehive, temperature of, 110, 155
Bees, bumble, 62, 155
  working, 62, 155
  mites in trachea of, 106
  temperature of, 110
  vibration of wings, 128
  specialized functions of, 154
  rhythmic activity of, 155-7
Begonia, buds of, 163
Beri-beri, 94
Bile, 91
Binary compounds, 21
Bird-lice, 62
Birds, food of, 71
  high temperature of, 109
  flight of, 129
  wing muscles of, 129
Black Cock, muscles of, 107
  speed of flight, 130
Blood, function of, 92
  oxygenated, 103
  corpuscles, 107
  amount of, in mammals, 109
Body-cavity, 88
Bombus lapidarius, hibernation of, 156
  B. terrestris, hibernation of, 156
Bonellia, 186
  male, 189
Bottom of sea, uniform conditions of, 58, 148, 149
Brachiopods, 35, 178
Bracken-fern, *Pteris*, 116, 171, 172
Branchiae, 103
Bryophyllum, 183
Buttercup, runners of, 163
  reproduction of, 187
Butterfly, 167

CABBAGE, Vitamin C in, 95
Calcium, in protein, 9
  in protoplasm, 41

Calories, 95
  number necessary, 96
Camel, food of, 79
  corpuscles of, 108
Capillaries, number of, in man, 110
  in horse, 110
  in guinea-pig, 110
Capillarity, 50
Capsule, 172
Carbo-hydrates, 22, 29, 40, 56, 57
  in vegetable food, 58
Carbon, in protein, 7
  in protoplasm, 41
Carbon dioxide, 21, 29, 32, 41, 99, 100, 101
  rate of oxidation controlled by, 102
  solubility in water, 102
  discharged, 110, 112, 113
Carboniferous age, 180
Caribou (American reindeer), 79
Carnation, 164
Carnivora, 81
Carnivorous animals, 34, 74, 51
  insects, 61, 62, 65
  fish, 67
  reptiles, 68-9
  birds, 71
Carotin, 93
Caspian Sea, 179
Cell-division, rhythm in, 135
Cells, 14, 17
  derivation of name, 17
  structure of cell walls, 18
  palisade, 53
  of sponge, 87
  movement of, 121
Cellulose, 18, 90
Centipedes, 47, 61
  respiratory system of, 104
Cestoda, 24
  larvae, movement of, 123
Cetacea, see Whales
Chaetae, 126
Chaetognatha, 178
Chaetopods, haemoglobin in, 107
Chaetopterus, 190
Chalk, 44
Chameleon-shrimp (*Hippolyte*), 135
Chara, 61, 157
Chimpanzee, 84
Chironomus, 107
Chiton, breeds at full moon, 146

INDEX

Chlorine, in protein, 7
Chlorocruorine, 108
Chlorophyll, 8, 22, 25, 28, 35, 36, 94
  absent in roots, 119
Chloroplast, 21, 29, 33
Chromosomes, 191
Chrysalides, shrinkage of, 65
Chyle, 91
Cicada, 46
Cilia, action of, 15, 17, 58, 116
  in frog, 16
  in mollusc, 16
  animals devoid of, 17
  in *Paramoecium*, 123
  of snail, 124
  rhythmic motion of, 134
Civilizations, ancient, disappearance of, 158
Clays, 43
Clepsina, 114
Clostridium, 37
Clover, Vitamin C in, 95
  on movement of, 120
Coccidium, 46
Coccus, 27
Cockroach, 61
Coelenterata, amoeboid cells in, 121
  habitat of, 178
  hermaphrodite, 185
Cold-blooded animals, temperature of, 110
Colonies of cells, 19, 164, 167
Communities, rhythmic activity of, 154–9
Compositae, reproduction of, 187
Coney, *Hyrax*, food of, 77, 78
Conjugation, 165, 189
Convoluta, influenced by tide, 144
Convolvulus, 118
Copepoda, 61, 66, 67
Cork, cells of, 17
Corpuscles red, number of, 107, 108
  size of, 108
  non-nucleated, 108
  white, 109
Cow, food of, 80
Crayfish, food of, 61
Crinoids (sea-lilies), food of, 59
Crocodiles, food of, 69
  mode of swimming, 127
Crocus, mode of growth, 137
Crows, 145

Crustacea, 26, 66, 169
  gills of, 103
  habitat of, 179
  spermatozoa of, 184
  hermaphrodite, 185
  parthenogenesis of, 187
Cuttings, reproduction by, 164
Cuttle-fish, *Loligo*, size of, 26
  attacks whales, 76
  backward movement of, 125
  habitat of, 178
Cyanea arctica, 26

Daddy-long-legs, *Tipula*, 47, 62
Daisy, 187
Daphnia, 107
  rise of, in water, 143
Darwin, on earthworms, 47, 59
Deer, food of, 79
  corpuscles of, 108
Denitrifying bacteria, 39, 42
Desman, *Myogale moschata*, 83, 84
Diarrhoea, infantile, 154
Diatoms, 25–6, 35, 36
  need light, 25, 148
  season for, 151, 153
Dictyota, influence of moon on, 147
Digestion, defined, 90–1
Dirt, in London air, 106
Divers, pearl, 111
Diving birds, 111
Dogfish, 67
  ova of, 176, 177, 183
Dolphins, 76
  in Cameroon River, 77
  habitat of, 179
Dragon-flies, 61
Drone larva, 62
Duck, flight of, 130
Duck-billed platypus, *Ornithorhynchus*, 72, 74
  low temperature of, 109
Dugong, 76, 77
Dust, in water injures fish, 105

Earthworms, work of, in raising soil, 47, 59
  respiratory organs in, 103, 114
  plasma in, 107
  movement of, 126

Earthworms (*continued*)
  reproduction of, 164
  monoecious, 185
Echinodermata, 178
Edentata, 74
Eel, migrations of, 129
  influence of moon on, 146
Egg, cells forming, 20
  fertilization of, 116
  rhythmic development of, 135
  contents of, 161
Eggs, as food, 58
  of fish, number of, 176
  shape and size, 182–3
  winter or resting, 187
Elater lineatus, 46 *n.*, 47, 140
Elephants, 77, 78
  corpuscles of, 108
Elodea, 162
Embryo sac, 172, 174, 187
Emulsification, 91
Endospores, 162
Enzymes, 94
Epidermis, 49, 53
Esquimaux, 86, 96
Euglenia viridis, 15 (fig. 4)
Excretion, 5

Faculties, decay of, 192
Fats, insoluble, 90
  animal, 94
  vegetable, 94
Fern, vegetative reproduction of, 164
  cilia of, 183
  see also under Bracken-fern
Fibro-vascular bundles, 51, 55 (fig. 17)
Fishes, food of, 66
  deep sea, 67
  mode of swimming, 126–7
Fission, 163, 189
Flaccidity of cells, 117
Flagella, vibratory process, 15, 116
Flame-cells, 100
Flat-worms, 182
Flea, jump of, 128
Fledglings, voracity of, 72
Flight, 128–31
Flounder, food of, 67
Flowering plants, parthenogenesis of, 187

Fluke, liver-, 169
  monoecious, 185
  alternation of generations, 189
Food, defined, 58
  necessary constituents of, 22
Food reserve, 33
Food-vacuole, 10
Foraminifera, 35, 178
Formaldehyde, 29, 31
Fox, food of, 82–3
  "flying," *see* Bats
Frog, food of, 68
  shortening of intestine, 69
  temperature of, 110
  temporarily anaerobic, 112
  heart, rhythmic action of, 137
  male, 178
  eggs of, 190
Frutti di mare, 141
Fungi, 22, 94
  reproduced by spores, 162, 164

Gall-flies, 188
Game-birds, speed of flight, 130–1
Gametes, 162, 166, 174
  movement of, 183
Gametophyte, 171, 172
Gas, natural, 34
  gangrene, 113
Gastric juice, 91, 97
Gastropoda, 107
Gel state in colloids, 122
Gemmation, 165
Gephyrea, 178
Geranium, dispersal of seed by, 121
Germ, 94
Gibbon, 83
Gills, of fishes, 103, 105
  of marine worms, 103
  of crustacea, 103, 105
Gill-slits, 103
Giraffe, food of, 80
Gland, salivary, 91
  of swift, 129
Globigerina, 14
Gloxinia, buds of, 163
Glucose, 29
Glycogen, 114, 131, 132
Goat, food of, 80
Gorilla, 84

Grampus, or killer-whale, food of, 76–7
Gregarines, parasites, 15
Grouse disease, 46
  food of, 71
  speed of, 130
Growing point, spiral movement of, 117, 136
Growth, a characteristic of living matter, 5
  influence of hormones on, 92
  influence of vitamins on, 95
  limit to, 160
Growth rings, absent in tropical trees, 147
Guard-cells, 53, 54 (fig. 16)
Guinea-pig, capillaries in muscles of, 110
Gulf Stream, effect of on salinity of sea, 150
Gymnosperms, 164

Haemocyanin, in crustacea, 108
Haemoglobin, 9, 36
  defined, 107
  distribution of, 107
  absent in crustacea, 108
Halobates, 179
Hare, 8, 107
Heart, muscles of, 108
  pulsations of, 111
  rhythmic action of, 138–40
  diminution of beats as age advances, 140
Hedgehog, food of, 83
Hellebore, 55 (fig. 17)
Herbivorous animals, 34, 74, 79, 81
  fishes, 67
Hermaphroditism, 185
Hermit-crab, change of sex of, 186
Herring, 67
  fisheries, influence of moon on, 146
Hippopotamus, food of, 79
Histolysis, 17
Holophytic types of feeding, 23
Holothurians (sea-cucumbers), 59
Holozoic type of feeding, 23
Hop, movement of, 118
Hormones, action of, 92
  derivation of name, 92
Horse, capillaries in muscles of, 110
Host, 15, 23

House-fly, mouth of, 63
  flight of, 128
Houseleek, buds of, 163
Humidity of atmosphere, 55
Humus, 39, 44
Hydra, 13, 164
  ova of, 183
Hydroid, 164
  Hydrozoa, 167, 189
Hyena, 81

Individuals, 20
Insectivora, 82
Insects, in soil, 46
  in manure, 47
  efficient respiration of, 103, 127, 155
  predominance of, explained, 104
  strength of, 128
  air-breathing, 179
Intestinal glands, rhythmic action of, 137
Intestine of man, length of, 89
Intussusception, 160
Ireland, reptiles in, 71
Iron, 9, 31
  in protoplasm, 41
  oxides, 43
  in haemoglobin, 107
Irritability of protoplasm, 12

Jackal, 81
Jelly-fish, *Medusa*, size of, 26
  food of, 58
  movement of, 123
  impart colour to sea, 151
  produced asexually, 168, 172
Jerboa, movement of, 128

Kangaroo, food of, 74
  movement of, 128
Katabolism, 58
Kelp, *Laminaria*, 170
Kidneys, 100
Kiwi, *Apteryx*, egg of, 183
Krogh, Dr, 110

Lacteals, 91
Lactic acid, 131–2
Lamprey, 167, 190

INDEX

Land-crabs, 180
Lankester, Sir E. Ray, 33
Larva, size of, 65
Larval stages, at sea level, 151
Laver, 125 n.
Lavoisier, 112
Layering, reproduction by, 164
Leaf mould, 37, 44
Leaves, number of, 34, 54–5
Leech, *Hirudo medicinalis*, 60, 61
  H. aulostoma, 61
  respiratory organs of, 103
  plasma in, 107
  temperature of, 110
  anaerobic, 114
  looping motion of, 126
  fertilization of, 182
  hermaphrodite, 185
Leguminosae, 38
Lemon juice, cure for scurvy, 95
Lemur, 86
Lepisma, 61
Life, definitions of, 1, 2
  Linnaeus's definition, 2
  originated in sea, 179
Lignin, 56
Limestones, sedimentary, 44
Liver, 91
  rhythmic action of, 137
Lizards, *Lacertae*, food of, 71
  L. vivipara, 71
  L. agilis, 71
Llamas, 79
Lobes, *see* Pseudopodia
London Pride, 32
Lump-sucker, male, 178
Lung-fish, *Dipnoi*, 179, 180
Lungs of vertebrates, 102
  function of, 103
  how protected, 103
  area of human, 111
  air in, 111
  rhythmic action of, 140

MAGNESIUM, in protein, 9
Malaria, 63
Malarial organism, 135
Malic acid, 119
Mammalia, fertilization of, 181
Mammals, milk of, 73
  in the sea, 76

Man, food of, 86, 96
  cooks food, 86
Manatee, 76, 77
Manure, effect of, 47
Margarine, 94
Marine worms, 59
  gills of, 103
  respiratory pigment of, 108
  reproduction of, 165
Marmot, 47
Marsupialia, food of, 74
Marsupium, 74
May-fly, *Ephemeridae*, 65, 167
Mealy-bugs, *Coccidae*, 188
Megasporangium, 174
Megaspores, 172
Mendeléeff, 35
Mesophyll, 53
Metabolism, 58. 92
Metaphyta, 13
Metazoa, 13, 87
Methyl orange, 30
Micropyle, 183
Microspores, 172
Mildews, 22
Milk, as food, 58
  of mammals, 73
Millipede, *Iulus terrestris*, 46, 47, 61
  movement of, 140
Mimosa pudica, 120
Mining and Factory Acts, 105
Mites, 17, 47
Mohl, Hugo von, 2, 7
Mole, 47, 88
Molluscs, 16, 18, 169
  feeding habits of, 58
  mode of movement of, 124, 125
  concentric lines of growth, 138
  habitat of, 178
  monoecious, 185
Mongoose, food of, 82
Monkeys, food of, 85
Monoecious plants, 185
Monotremata, young of, 73
  food of, 74
Moon, influence on
  marine invertebrates, 141–2, 145, 146
  molluscs, 144, 146
  birds, 145
  fish, 146
  plants, 147

200 INDEX

Mosquito, female, 19, 64
  buzzing of, 127
  flight of, 129
  periodic activity of, 144
Moss, 172
  club-moss, 164
Moth, clothes, 64
  flour, 64
Moulds, 22
Mouse, food of, 81
Movement, of plants, 114–22
  of plants, how caused, 117
  sleep, 119
  amoeboid, 121
  of animals, 122
Mulberry, 164
Muscles, contraction of, 131
  relaxation of, 131
Mushrooms, 22
Musk-ox, *Ovibos moschatus*, 78, 80
Musquash, 81
Mussel, 16, 58
  lines of concentric growth, 139 (fig. 49)
Myogenic theory, meaning of, 138
Myriapoda, 140
Myxomycetes, 9
μ, defined, 10 n.

N*EMATODA*, *see* Roundworms
Nemertine worm, 108
Nephelis, 114
Nettle, stinging, 187
Newts, food of, 68
Nitrates, 33, 40, 89–90
Nitrites, 39
Nitrobacter, 39
Nitrogen, in protein, 7
  in protoplasm, 41
Nitrogen compounds, sources of, 37
Nitrogen cycle, 37, 42
Nitrosomonas, 39
Nitrous acid, 39
Noctiluca, rhythmic brightness of, 135
Nodules, 38
Nucellus, 172, 174
Nucleus, 10
  of blood corpuscles, 108

O*DONTISYLLIS ENOPLA*, 146
Off-spring, number of, 175

Oikomonas, rhythmic action of, 143
Oikopleura, *see* Ascidians
Oil, mineral, organic origin of, 34–5
Old age, 191
Operculum (gill-cover), 104
Opossum, food of, 74
Orange-juice, 95
Orange tree, independent of seasons, 148
Orang-utan, *Simia satyrus*, food of, 85, 86
Orchids, number of seeds, 175
Orthonectid, 167
Osmosis, 48–9, 54, 55
Ostracoda, scavengers of sea, 66
  spermatozoa of, 184
  parthenogenesis of, 187
Ostrich, egg of, 183, 184
Otolith, concentric rings in, 138
Oxygen, 7, 12, 28, 29, 99, 101, 110
  tension, 102, 114
  absorbed by muscles, 131
Oysters, Cicero on, 141
  change of sex of, 185

P*ALOLO WORM*, *Eunice*, spawning habits of, 145
  reproduction of, 164
Pancreas, 91, 92
  rhythmic action of, 137
Pancreatic juice, 91
Pap (bee), 62, 87
Paramoecium, 10, 123
  rhythmic motion of cells in, 133
  ciliary action of, 134
  conjugation of, 165, 189
  fission of, 190
Parasites, 18, 23, 24
Parthenogenesis, 186–91
  artificial, 190–1
Pasteur, 112
Pearl, rhythmic growth of, 138, 139
Pecten, mode of swimming, 124
Pellagra, 94, 95
Perennial plants, 147
Peripatus, 182
Perissodactyla, 77
  food of, 78
Peristalsis, meaning of, 137
Periwinkles, influenced by tide, 144
Petrels, 72

INDEX

Phagocytes, 109, 167
Phalanger, 74
Phosphorescence of sea, 135, 146
  greatest in July, 151
Phosphorus, in protein, 9
  in protoplasm, 41
Photo-synthesis, 31, 56, 93
Phyllopoda, 187
Phylloxera, 188
Phytoplankton, 26
Picrines, 36
Pig, food of, 79
Pigment cells, rhythm in, 135
Pigments, animal, 36
Pilchard, young of, 67
Pituitary body, 190
Plaice, food of, 67
Plankton, 151, 153
Planorbis, 107
Plant-cells, 17
Plant lice, 63
  reproduction of, 188
Plantain, seeds of, 175
Plantain-eaters, 36
Plasma, 107, 110
Platanus orientalis, 51 (fig. 14)
Plover, American Golden, migration of, 129–30
Pluteus, 190
Pollen sac, 172
Pollen tube, 174
Polychaeta, 59
Polypodium vulgare, 171
Pontecypris, 184
Pore-space, 44
Pores, ambulacral, 105
Porpoises, 76
Potassium, in protein, 9
  in protoplasm, 41
Potato "eyes," 163
Prairie dog, 47
Proteins, composition of, 7, 9, 29, 36, 89
  molecules of, 8
  in plants, 29, 37
  in animal food, 56, 58
  insoluble in water, 90
Prothallus, 171, 174
Protophytes, 166, 167
Protoplasm, defined, 2
  motile character of, 3
  constitution of, 7
  characteristic of, 36

Protoplasm (*continued*)
  cycle, 41
  oxidation of, 100
Protoplasmic egg, 183
Protozoa, 45, 59, 142
  immortal, 166, 167
Pseudomonas, 38
Pseudopodia, 10, 11, 12, 122, 123
Pteropoda, 76, 178
Pupae, shrinkage of, 65

Rabbit, 81
Radiolarians, 178
Radish, seeds of, 175
Rat, 81
  fed on food lacking Vitamin A, 94
Red Indians, 96
Red Sea, 141
Reindeer, food of, 79
Reproduction, of protoplasm, 5, 12
  when limit of growth is reached, 160
  vegetative, 162, 164
Reptilia, eggs of, 181
Respiration, 33, 92
  of plants, 99–101
  of insects, 103
Respiratory action of protoplasm, 4
  organs in animals, 101–5
  essential feature of, 103
Reticulate sap vessels, 51
Retractor bulbi, 68
Rhinoceros, destroys vegetation, 79
Rhizocephala, change of sex, 185
Rhizome, 115, 116
Rhythm, 5, 6, 133–59
  in cells, 134–6
  in tissues, 136–8
  in organs, 138
Rickets, how cured, 93
Rings, annual, 136, 138
Rodents, food of, 81
Root-hairs, 48
Root pressure, 52
Rose, buds of, 163
  reproduction of, 187
Rotifers, 16
  parthenogenetic, 187
Roundworms, *Nematoda*, 46, 47, 113
  movement of, 123–4
  spermatozoa of, 184
  sexual condition, 186

INDEX

"Royal jelly," 62, 87
Rumination, 80-1.

Sacculina, 186
Saliva, 91
Salmon, 179
Salpa, 169
Salts, in soil, 49
 in food, 57
Sand, in water, 105
Sands, 43
Sap, ascent of, 49
 vessels, 49, 50
 rate of ascent, 52
Saprophytes, 23, 27
Sarcina, 27
Sardines, food of, 87
Sargasso Sea, 179
Saw-flies, *Tenthredinidae*, 188
Scalariform vessels, 50
Scales (fish), concentric rings in, 138,
Scarabeus sacer, 64
Scarlet runner, movement of, 118
Scorpions, food of, 65
Scurvy, causes of, 93, 94
Scurvy-grass, *Cochlearia*, 95
Sea-anemones, 58
 movement of, 123
 reproduction of, 164
Sea-cow, *Sirenia*, 77
Sea-horse, *Hippocampus*, 108, 178
Sea-lilies, *Antedon*, 59, 105, 178
Sea-snakes, 179
Sea-urchin, movement of, 125, 141
 ovaries of, 141
 Diadema setosa, rhythmic growth of generative organs of, 141
 habitat of, 178
 eggs of, 190
Sea-water, 95
Seals, *Carnivora*, 76, 77, 83
Seaweed, brown, 169
Secretin, 92
Secretion by living plants and animals, 4-5
Sedge, rhizome of, 116-7
Seed-bearing plants, 175
Sepia, 182
Sessile organism, 16, 22, 115, 122, 169
Sharks, 67
 basking shark, *Cetorhinus*, 67

Sheep, food of, 80
Shepherd's purse, seeds of, 175
Shrew, 83
 water-, 83
Sieve-plates, 51
 -tubes, 51, 55
Silica, 43
Silk-worm moth, eggs of, 190
Silts, 43
Skate, number of eggs, 177
Skeletal elements of cells, 56
Skeletons, flinty, of diatoms, 25
Skunk, 82
Sloths, food of, 75-6
Slow worm, *Anguis fragilis*, 71
Slugs, rate of movement, 125
Smelt ("Grunion"), spawning habits of, 142
Snail, mode of movement of, 124
 rate of movement of, 125
 mode of reproduction, 182
Snakes, method of feeding, 70
 poison of, 71
Sodium, in protein, 9
 in protoplasm, 41
Soil, composition of, 43
 insects in, 46
Sol state in colloids, 122
Solen, 107
Soma, 166
Spallanzani, 112
Species, number of, 104
Specific gravity, of animals, 173
 of plants, 173
Spergula arvensis, size of stomata, 54
Spermaceti, oil from, 76
Spermatophores, 182
Spermatozoa, of fish, 176, 182
 of mammals, 181
 larger and smaller, 184
Spermatozoon, 15, 16, 20, 116, 166
 size and shape, 184
Spice trade, importance of, 97
Spiders, food of, 65
 respiratory system of, 104
 generation of, 169
Spirogyra, 29
 growth of, 137
Sponge, gemmules of, 165
 horny, 178
Sponges, 87, 185

INDEX

Sporangia, 171
Spores in yeast, 161
 in unicellular plants, 165
S orophyte, 171, 172
Springtails, 127
Squirrel, 81
 flying, 81
Squirting cucumber, dispersal of seeds by, 121
Starch, 31
 insoluble in water, 90
Starch-grains, concentric rings in, 138
Starfish, 67
 movement of, 125, 126
 habitat of, 178
Statoblasts of Polyzoa, 165
Stickleback's nest, 177
Stigma, 174
Stigmata (or Spiracles) in insects, 104, 105
Stomata, 32, 53, 55
 number of, 54
Strawberry, runners of, 163
Sugar, 12, 29, 31, 33, 38
 cane, 33, 164
 grape, 33
Sulphur, in protein, 7
Sundew, movement of, 119
Sunflower, section of stem, 50
Sunlight, effect of on life in sea, 145–53
Swallows, migration of, 129
Swan, speed of flight of, 130
Swift, flight of, 129, 131
Syllis ramosa, 165

Tadpole, food of, 69
 development of, 180
Tapeworms, 24, 46, 77, 113, 169, 185
Tapir, food of, 77, 79
Telegraph plant, 117
 rotatory leaves of, 136
Temperature, of plants, 101
 normal of human body, 109
 of birds, 109
 of cold-blooded animals, 110
 of sea, 150–3
 of soil, 153–4
Tendrils, 118

Tenebrio, 64
Termites (white ants), 62–3
 specialized functions of, 154
Ternary compounds, 21
Tetanus (lock-jaw), 113
Thorny-headed roundworms, 24
Thrush, food of, 72
Thyme, 187
Thyroxin, 92
Tides, influence on marine creatures of, 144
 rhythm of, 149
Tinea moth, 63
Tissues, defined, 13
 action of, 110
Toads, food of, 68
Tobacco plant, seeds of, 175
Tortoises, food of, 69
 movement of, 127
 rhythmic action of heart, 137
Tortrix viridana, destroys oak leaves, 65
Tracheae, 104, 105
 bladders in, 105
Tracheids, 49, 50, 55
Trematoda (flukes), 87
 larvae, movement of, 123
Trench fever, 63
Tropic movements, 119
 geotropic, 119
 heliotropic, 119
Tropics, slight seasonal changes in, 147
Tubers, 56
Turacin, 36
Turbellaria, 87, 123
 amoeboid cells in, 121
 gliding movement of, 123
 fertilization of, 182
 monoecious, 185, 186
Turbot, number of eggs, 176
Turgidity of cells, 56, 117, 120, 121
Turtles, snapper, 69
 green, 69
 logger-headed, 69
 leathery, 69
 mode of swimming, 127
 habitat of, 180
Twining stems, 118
Typhoid fever, 63
Typhus fever, 63

INDEX

UNGULATA, 77
Unicellular organisms, 14–15, 45
  rhythmic motion of, 47, 134
  shape of, 160
  reproduction of, 161
Urea, excretion of, 5, 8, 40, 100

VACUOLE, contractile, 11, 100
  of Infusoria, 133
  intervals of contraction, 133
Valves, in bats' wings, 138
Vascular bundles, 51, 55 (fig. 17)
Vegetable foods, 58
Vegetarians, 96
Vegetative cell, 174
Veins, of leaves, 54
Venus' Fly-trap, movement of, 119
Vertebrae (fish), concentric rings in, 138
Vertebrata, food of, 66
  lungs of, 102
  not dominant species, 154
  reproduction of, 181
Vespa, 155
Vesuvius, eruption of (1906), 105
Vibration of insects' wings, 103, 128
Vibrio, 27
Viper, temperature of, 110
Vitamin A, B, C, 93, 94, 95
Vitamins, 57, 92
  essential to life, 93
Volvox aureus, 19 (fig 6)

WALLABY, 74
Walrus, 83
Wapiti, 79

Washington elm, 34
Wasps, 62
  specialized functions of, 154–5
  rhythmic activity of, 155, 157
  importance of queen, 158
Water, loss of, in transpiration, 54–5
  in food, 57
Water-worms, movement of, 126
Weevils, destruction by, 65
Westminster Hall, 64
Whales, *Cetacea*, whale-bone, food of, 67, 76
  toothed, food of, 76
  spermaceti, 76
  killer, 76–7
  breathe in water, 111
Wire-worm, *see* Millipede and *Elater*
Wolf, Tasmanian, 74
Wombat, 74
Woodlouse, 47

XANTHARPYIA COLLARIS, 84, 85
Xestobium tessellatum (death-watch beetle), 64

YEAST CELLS, 112
  division of, 161

ZOOIDS, 164
  asexual reproduction of, 167
Zoospores, 162

www.ingramcontent.com/pod-product-compliance
Ingram Content Group UK Ltd.
Pitfield, Milton Keynes, MK11 3LW, UK
UKHW040656180125
453697UK00010B/202